Design Science and Innovation

Series Editor

Amaresh Chakrabarti, Centre for Product Design and Manufacturing, Indian Institute of Science, Bengaluru, India

The book series is intended to provide a platform for disseminating knowledge in all areas of design science and innovation, and is intended for all stakeholders in design and innovation, e.g. educators, researchers, practitioners, policy makers and students of design and innovation. With leading international experts as members of its editorial board, the series aims to disseminate knowledge that combines academic rigour and practical relevance in this area of crucial importance to the society.

Ajita D. Singh · Rauf Iqbal · Vivek Khanzode
Editors

Sports Ergonomics

Proceedings of HWWE 2022

Springer

Editors
Ajita D. Singh
Department of Sports Science
Punjabi University Patiala
Patiala, Punjab, India

Rauf Iqbal
Ergonomics Lab
National Institute of Industrial Engineering (NITIE)
Mumbai, Maharashtra, India

Vivek Khanzode
National Institute of Industrial Engineering (NITIE)
Mumbai, Maharashtra, India

ISSN 2509-5986 ISSN 2509-5994 (electronic)
Design Science and Innovation
ISBN 978-981-97-7803-4 ISBN 978-981-97-7804-1 (eBook)
https://doi.org/10.1007/978-981-97-7804-1

This Springer imprint is published by the registered company Springer Nature Singapore Pte Ltd.
The registered company address is: 152 Beach Road, #21-01/04 Gateway East, Singapore 189721, Singapore

Preface

This book is the proceeding of International Ergonomics conference in the name of "Humanising Work and work environment" which is endorsed by IEA (International Ergonomics Association) organised by Punjabi University, Patiala. The conference proceedings present a variety of issues confronted by researchers and practitioners in the field of sports ergonomics. The book highlights the latest research on various human factors and ergonomics in sports and provides practical guidance on a range of methods for describing, representing and evaluating human performance in sports domains. The book will be useful for the researchers, professionals and consultants in the field of sports ergonomics, sports analytics and biomechanics.

Patiala, India	Ajita D. Singh
Mumbai, India	Rauf Iqbal
Mumbai, India	Vivek Khanzode

Contents

Editors and Contributors

About the Editors

Dr. Ajita D. Singh is teacher, researcher, and exercise physiologist. She is Professor and former Head Department of Sports Science, Faculty of Medicine, Punjabi University, Patiala. She holds Ph.D. degree in Exercise Physiology. Her main research interests are exercise physiology, ergonomics and human factors, sports ergonomics, biomechanics of human movement, occupational ergonomics and safety management. She is actively involved in extensive research activities and published in national and international peer reviewed journals. She is actively involved in renowned national and international societies like Punjab Academy of Sciences, Nutritional Society of India, Indian Society of Ergonomics, Physiological Society of India, and Physical Education Foundation of India. She contributes as Associate Editor and Reviewer for several international journals. She has presented her research as invited speaker in various national and international conferences in India and abroad. She has guided six Ph.D. students and 14 students are still registered under her guidance.

Dr. Rauf Iqbal is Professor of Ergonomics and Human Factors Engineering at the National Institute of Industrial Engineering (NITIE), Mumbai. Currently, he is Dean-Student Affairs of NITIE. His area of interest includes work system design, engineering anthropometry and workspace design, biomechanics of human movement, occupational ergonomics, and safety management. Dr. Iqbal has published more than 125 as journal articles, edited books, book chapters, and conference proceedings. He has guided ten Ph.D. scholars. He is member of various international and national committees, such as Foundation of Professional Ergonomics, USA, BRICS Plus Executive Committee for Human Factors and Ergonomics, Council Member of Asian Council on Ergonomics and Design, Ergonomics Sectional Committee for Bureau of Indian Standards, and General Secretary of Indian Society of Ergonomics. He has carried out various projects funded by the Government of India and International Labour Organization. He conducts training and consultancy for numerous industries in the areas of ergonomics, work study, work systems design, and safety management.

Prof. Vivek Khanzode is a Professor of Industrial Engineering and Manufacturing Systems at the National Institute of Industrial Engineering (NITIE), Mumbai. He completed M. Tech. and Ph.D. in Industrial Engineering from IIT Kharagpur. His area of interest includes work system design, manufacturing system engineering and design, lean manufacturing, simulation modeling, and workplace ergonomics. Dr. Khanzode is involved in training programs and consultancy assignments for corporates and public sector undertakings in the field of operations excellence, productivity and efficiency improvement, process capacity and bottleneck analysis, and simulation modeling in manufacturing, service, and project operations. He is the member of Indian Society of Ergonomics and associated with HWWE conference for the past decade.

Contributors

Shumetie Agonafer Babu Sport Academy, Department of Sports Science, Debre Markos University, Debre Markos, Ethiopia

Neepa Banerjee Bhairab Ganguly College Affiliated to West Bengal State University, Kolkata, India

Satabdi Bhattacharjee University of Calcutta, Kolkata, India

Priyanka Biswas Department of Food and Nutrition, PHAU, West Bengal State University, Kolkata, India

Enakshi Chakraborty Department of Food and Nutrition, PHAU, West Bengal State University, Kolkata, India

Sandipan Chatterjee Department of Food and Nutrition, PHAU, West Bengal State University, Kolkata, India

Arpan Chattopadhyay Department of Sports Bioscience, School of Sports Sciences, Yoga and Education, Central University of Rajasthan, Ajmer, Rajasthan, India

Subhasmita Das Department of Sports Bioscience, School of Sports Sciences, Yoga and Education, Central University of Rajasthan, Ajmer, Rajasthan, India

Nishan Singh Deol Department of Physical Education, Punjabi University, Patiala, India

Tirthankar Ghosh Ergonomics and Exercise Physiology Laboratory, Faculty of Health and Wellness, Sri Sri University, GodiSahi, Cuttack, Odisha, India

Sweta Goyal Department of Home Science, Kurukshetra University, Kurukshetra, Haryana, India

Ruhi Grewal Department of Home Science, Kurukshetra University, Kurukshetra, India

B. Hemanth Naick Department of Sports Bioscience, School of Sports Sciences, Yoga and Education, Central University of Rajasthan, Ajmer, Rajasthan, India

Ekta Kapri Department of Family Resource Management, CCSHAU, Hisar, Haryana, India

B. R. Kasvan Department of Sports Bioscience, Central University of Rajasthan, Ajmer, India

Tarvinder Jeet Kaur Department of Home Science, Kurukshetra University, Kurukshetra, Haryana, India

Harish Kumar Department of Sports Science, Punjabi University, Patiala, Punjab, India

Deepak Kumar Mallick Ergonomics and Exercise Physiology Laboratory, Faculty of Health and Wellness, Sri Sri University, GodiSahi, Cuttack, Odisha, India

Manju Mehta Department of Family Resource Management, CCSHAU, Hisar, Haryana, India

Bikram Moyra Department of Sports Bioscience, School of Sports Sciences, Yoga and Education, Central University of Rajasthan, Ajmer, Rajasthan, India

Shankarashis Mukherjee Department of Food and Nutrition, PHAU, West Bengal State University, Kolkata, India

Arvind Pareek Department of Botany, Maharshi Dayanand Saraswati University, Ajmer, India

Sohini Paul Ergonomics and Exercise Physiology Laboratory, Faculty of Health and Wellness, Sri Sri University, GodiSahi, Cuttack, Odisha, India

Anilendu Pramanik Department of Sports Science, Punjabi University, Patiala, Punjab, India

Sunil G. Purohit Department of Sports Bioscience, Central University of Rajasthan, Ajmer, India

Paladi Ramya Department of Sports Bioscience, School of Sports Sciences, Yoga and Education, Central University of Rajasthan, Ajmer, Rajasthan, India

Divik Ranjan Department of Sports Bioscience, School of Sports Sciences, Yoga and Education, Central University of Rajasthan, Ajmer, Rajasthan, India

Tanaya Santra Department of Food and Nutrition, PHAU, West Bengal State University, Kolkata, India

Ajita Singh Head Department of Sports Science, Faculty of Medicine, Punjabi University, Patiala, India

Kiran Singh Department of Family Resource Management, CCSHAU, Hisar, Haryana, India

Malook Singh Department of Sports Science, Punjabi University Patiala, Patiala, India

Manjeet Singh Department of Sports Science, Punjabi University Patiala, Patiala, India

Neha Singh Department of Sports Bioscience, Central University of Rajasthan, Ajmer, India

Paramvir Singh Department of Sports Science, Punjabi University Patiala, Patiala, India

Harshita Tak Department of Sports Bioscience, School of Sports Sciences, Yoga and Education, Central University of Rajasthan, Ajmer, Rajasthan, India

Ridham Thakral Department of Sports Science, Punjabi University, Patiala, Punjab, India

Madhvee Tiwari MBS Punjab Sports University, Patiala, India

Rahul Tiwari Sports Authority of India, NS NIS, Patiala, India

Anjali Vashishth Department of Sports Bioscience, Central University of Rajasthan, Ajmer, India

Naseer Ud Din Waza Department of Physical Education, Punjabi University, Patiala, India

Chapter 1
A Study on Physical Fitness and Cognitive Status of Adult Bengalee Males Undergoing Training in Football

Satabdi Bhattacharjee, Tanaya Santra, Enakshi Chakraborty, Priyanka Biswas, Sandipan Chatterjee, Neepa Banerjee, and Shankarashis Mukherjee

1 Introduction

Football is the most commonly played team sports that characterized as highly dynamic and complex interaction of strategic, technical, and tactical dimensions at team level; on the other hand, physical, physiological, and technical dimension is required at individual level (Carling and Orhant 2010). It is a complex interplay between a moderate to high-intensity intermittent actions, where players have to perform difficult and unpredictable movement patterns depending on the game situations. Therefore, football players require to develop different aspects of fitness including repeated sprinting ability, acceleration and deceleration ability, aerobic and anaerobic endurance capability and also need to perform short bouts of low to moderate intensity actions (Bradley et al. 2009). Fitness rather physical fitness is categorized in two parts, i.e., health-related and performance- or skill-related physical fitness. As the name suggests, health-related fitness involves mainly physical fitness associated with the body physique, composition, muscular endurance, strength, and flexibility. On the other way, performance-related fitness includes specific skills especially physical and mental coordination involving speed, reaction time, agility, balance, cognitive ability, cognitive flexibility, metacognition, and decision-making that are suggested to be crucial for athletic performance because at the time of playing

S. Bhattacharjee
University of Calcutta, Kolkata, India

T. Santra · E. Chakraborty · P. Biswas · S. Chatterjee · S. Mukherjee (✉)
Department of Food and Nutrition, PHAU, West Bengal State University, Kolkata 700126, India
e-mail: msasish@yahoo.co.in; drsmlab@gmail.com

N. Banerjee
Bhairab Ganguly College Affiliated to West Bengal State University, Kolkata, India

A. D. Singh et al. (eds.), *Sports Ergonomics*, Design Science and Innovation,
https://doi.org/10.1007/978-981-97-7804-1_1

a player has to respond to numerous stimuli, as they need to take decision with no time during a match (Chen et al. 2002; Raylor and Donovan 2013) and the pre-frontal cortex is mainly responsible for that. In this background, a work was carried out to assess the impact of being trained in football on physical fitness and cognitive status of Bengalee adult males.

2 Methodology

Many coaching centers in Hooghly district, West Bengal, giving training to male individuals on football were contacted, and the requirements for the study was also communicated to the concerned authority. Only the centers that had given prior permission for carrying out the study were selected. The measurements were obtained in the morning by the same individual from 35 arbitrarily chosen volunteered adult male Bangalee individuals of 18–22 years (age range) receiving training in football with a minimum experience of playing football for at least six years and practicing it about six days a week with a duration of two hours without having any chronic disease history (self-reported); they basically formed the Football Practicing Group (FG). At first, very basic information like their playing experience (years), regular playing time (hour), playing position, educational qualification, and socioeconomic status was recorded in a pre-designed schedule. Stature (m) (to the nearest 0.1 cm) using anthropometric measurement set, body weight (kg) (to the nearest 0.1 kg) using an electronic weighing scale with individuals without shoes and in socially acceptable lightweight exercising clothes were measured, and BMI (kg m^{-2}) was calculated. Percentage of body fat of the subjects was calculated after obtaining standardized skin fold measurements from specific sites (Jackson and Pollock 1978). Sprinting ability of the subjects assessed by 50-yard dash (Subramanya and Pasodi 2011). To measure explosive leg power over a horizontal distance of the subjects, standing broad jump test (SBJT) (Bianco et al. 2015) conducted. Lower limb power of the individuals measured using vertical jump test (VJT) (Kale et al. 2009). Trail making test (TMT) (Vestberg et al. 2012) and D-KEFS design fluency test (DFT) (Suchy et al. 2010) are conducted to assess cognitive ability and cognitive flexibility of the individuals. In the same manner, assessments were also carried out on 40 adult Bengalee individuals of comparable age, similar socioeconomic and ethnic background without formal training in any form of exercise including football; they constituted the control group (CG). After getting the data, statistical analyses carried out and < 0.05 was chosen as the level of significance (P).

3 Results

Obtained data is graphically presented.

The physiological and physical characteristics of the study participants (FG and CG individuals) are presented in Table 1.

Body composition status of the individuals was estimated using BF% and BMI (kg m^{-2}) and presented in Fig. 1a, b. It is observed that FG individuals have significantly ($P < 0.05$) lower BF% and BMI (kg m^{-2}) than CG males.

Physical fitness status of the study participants has been estimated using vertical jump test, 50-yard dash, and standing broad jump test and presented in Fig. 2a–c. It has been found from the results that FG individuals have significantly better physical fitness status compared to their CG counterparts.

Cognitive ability of the study participants has been assessed using trail making test (TMT), consisting of TMT A and TMT B. In Fig. 3a, b, trail making test scores of the study participants have been presented. It has been found from the test results that the cognitive ability and cognitive flexibility of the FG individuals is significantly ($P < 0.05$) better compared to the CG counterparts.

Table 1 Physiological and basic physical profile of study participants

Variables	FG	CG
Age (years)^	19.1 ± 1.23	19.5 ± 0.70
Body weight (kg)*	54.7 ± 8.32	88.9 ± 10.94
Systolic blood pressure (mmHg)^ $_{\text{pre-exercise}}$	115.3 ± 12.70	119.1 ± 12.69
Diastolic blood pressure (mmHg)* $_{\text{pre-exercise}}$	67.9 ± 10.02	88.0 ± 6.80
Heart rate $_{\text{pre exercise}}$ (bpm)*	65.1 ± 9.95	84.7 ± 9.41

* $P < 0.05$, ^ns, AM ± SD

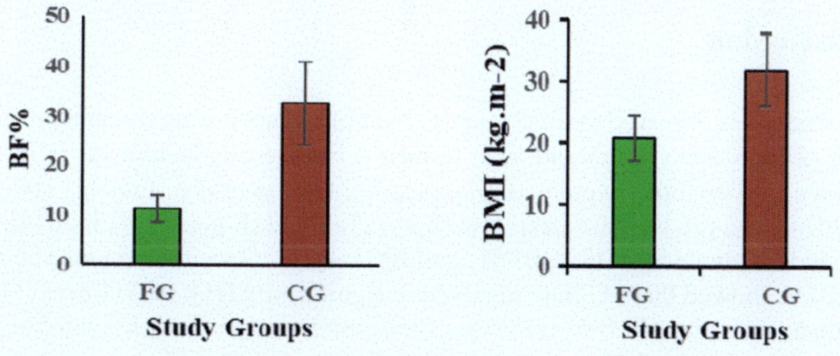

(a) Body Fat %*

(b) BMI (kg.m^{-2})*

Fig. 1 Body fat % and BMI compared between FG and CG males

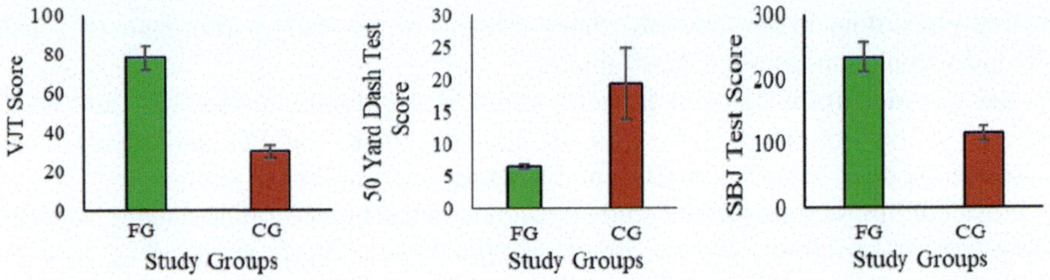

(a) VJT (cm)* (b) 50 Yard Dash (sec)* (c) SBJ (cm)*

Fig. 2 Physical fitness parameters in terms of VJT, 50-yard dash, and SBJ compared between FG and CG males

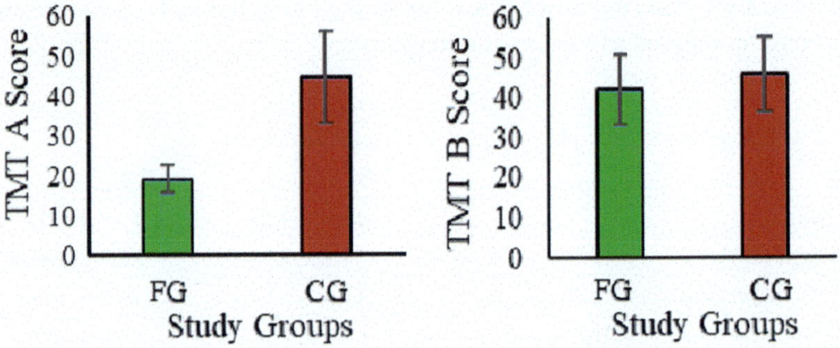

(a) TMT A (sec)* (b) TMT B (sec)*

Fig. 3 Cognitive ability in terms of trail making test score compared between FG and CG males

4 Discussion

In this study, it is observed that FG and CG individuals are not significantly differing in case of their age (years) and SBP (mmHg), but there is significant ($P < 0.05$) difference in their body weight (kg), diastolic blood pressure (mmHg) (Bangsbo et al. 2014), and pre-exercise heart rate (bpm) (Table 1) (Bangsbo et al. 2014).

To study body composition, BF %, and BMI (kg m^{-2}) were selected. The result of the BF% showed that FG individuals have significantly ($P < 0.05$) lower (11.3%) BF% than CG males (32.6%) (Fig. 1a). The result showed similarity with the prior work conducted on football players of English Premier League, Japan, and Zimbabwe where it has been observed that body fat % varies from 9.9–12.9% (Sutton et al. 2009), 8.5–13.7 % (Tahara et al. 2006), and 9.2–11.2 % (Masocha and Katanha 2014), respectively. The finding is also in agreement with the past work carried out on trainee Indian classical danseuses (Mukherjee et al. 2012, 2014, 2014b; Banerjee et al. 2014a, 2014b) and sub-division adolescent Bengalee football players (Bhattacharjee et al. 2016, 2015a, 2015b; Bhattacharje et al. 2018). BMI is also known as a threat for developing of different diseases including obesity and other health issues. In the present study, it is observed that the BMI (kg m^{-2}) is significantly ($P < 0.05$) lower in FG males compared to the CG males (Fig. 1b), and this finding is also similar with our prior work carried out on Indian classical dancers (Chaterjee et al. 2014; Kundu et al. 1970).

To study physical fitness of the study participants, vertical jump test, 50-yard dash, and standing broad jump test were selected. The lower limb power of the individuals was estimated by vertical jump test. The distance that is covered during vertical jump is having a straight association with the force generated by lower limb muscle fibers, made up of muscle proteins. It is observed that FG males are having significantly ($P < 0.05$) superior lower limb power and power output than CG males (Fig. 2a), indicating better conditioned lower muscle mass responsible for generating power that is required during a 90 min football match. The finding is in agreement with our past work carried out on adult Bengalee footballers (Bhattacharjee et al. 2021a). Sprinting ability is one of the important aspects in football as it is related with various dynamic actions carried out during resistance created by the opponent during a match Negra et al. (n.d.). In this work, FG individuals have showed significantly ($P < 0.05$) better sprinting ability estimated by 50-yard dash test compared to the CG males (Fig. 2b). The result is in consonance with our previous work conducted on adult Bengalee male football players (Bhattacharjee et al. 2021a).

Football is a game which has been characterized as an intermittent sport requiring both the aerobic and anaerobic power. Football is a game that needs 70% aerobic endurance and 30% anaerobic power. Anaerobic power is a critical element, and it is required mainly during short-term sprinting actions, explosive efforts during a match. High-intensity explosive actions are lasting in football for about 8 seconds, and most importantly the energy pathway which is responsible for that is known as ATP and creatine phosphate system. In the present work, the explosive leg power of the individuals has been measured using standing broad jump test (Krishnan et al. 2017), and it is found that FG males have showed significantly ($P < 0.05$) better score in standing broad jump test compared to their CG counterparts (Fig. 2c).

The cognitive skills of the subjects were assessed by trail making test and design fluency test. It is found that FG individuals have scored significantly ($P < 0.05$) better in both the cognitive ability test assessed by trail making test (Fig. 3) and cognitive flexibility test assessed by design fluency test (Fig. 4) than CG males. Modern research has been found a relation between physical fitness and cognitive functioning. The pre-frontal cortex is dealing with this. It is reported in studies that consistent practice of a particular physical activity for a long period of time with particular intensity, modality, and duration, may bring a positive influence on cognitive domain. The findings are in agreement with the prior work carried out on adult Bengalee football players (Bhattacharjee et al. 2021b).

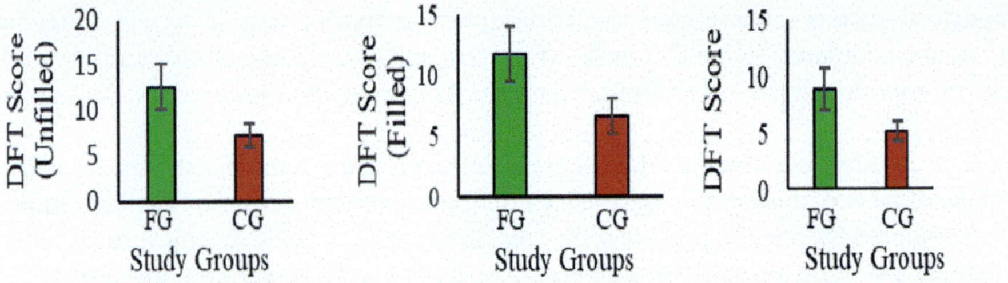

(a) DFT A (count in numbers)* (b) DFT B(count in numbers)* (c) DFT C (count in numbers)*

Fig. 4 Cognitive flexibility in terms of design fluency test score compared between FG and CG male

5 Conclusion

It may be summarized from the result of the present work that adult Bengalee males receiving regular training in football for minimum 6 years and practicing it regularly at least a period of 6 days a week and a minimum duration of two hours have a significantly ($P < 0.05$) better physical fitness along with cognitive status than CG males.

Acknowledgements Authors are thankful to the participants and the concern authorities for their continuous support and cooperation.

References

Banerjee N, Chatterjee S, Kundu S, Bhattacharjee S, Mukherjee S (2014b) Effect of regular practicing Bharatnatyam dancing exercise on body fat of urban female teenagers. Indian J Clin Anat Physiol 1(1):29–33. ISSN: 2394-2118 Print, 2394–2126 Online)

Banerjee N, Chatterjee S, Chatterjee A, Kundu S, De S, Mukherjee S (2014a) Relationship between two body adiposity estimating methods: a study in Bengalee female adolescents being trained in Bharatnatyam dance. Int Physiol 2(2):41–49 ISSN 2347-1506

Bangsbo J, Junge A, Dvorak J, Krustrup P (2014) Executive summary: football for health—prevention and treatment of non-communicable diseases across the life span through football. Scand J Med Sci Sports 24(1):147–150. https://doi.org/10.1111/sms.12271

Bhattacharje S, Chatterjee A, Banerjee N, Chatterjee S, Mukherjee S (2018) Customized helmets for two-wheeler users: an anthropometric study. Indian J Biol Sci 24:45–50. ISSN: 0972-8503

Bhattacharjee S, Banerjee N, Chatterjee S, Chatterjee S, Chatterjee A, De S, Mukherjee S (2015) Impact of receiving training in football on body physique and fitness status of adult Bengalee males. In: Proceedings of humanizing work and working environment, caring for people, pp 404–410. ISBN: 978-93-5258-836-7

Bhattacharjee S, Banerjee N, Chatterjee S, Chatterjee A, De S, Mukherjee S (2015) Impact of receiving training in football on obesity and physical fitness status: a study in Bengali male adolescents. IP 3(1):41–47 ISSN: 2347-1505. https://doi.org/10.21088/ip.2347.1506.3115.6

Bhattacharjee S, Banerjee N, Santra T, Chatterjee S, Chatterjee S, Mukherjee S (2016) Body physique and motor ability status of bengalee adult males receiving training in football. Indian J Biol Sci 22:1–9. ISSN: 0972-8503

Bhattacharjee S, Santra T, Chatterjee S, Biswas P, Banerjee N, Mukherjee S (2021a) A study on flexibility and fitness status of adult Bengalee males undergoing training in football. In: Proceedings of humanizing work and working environment, ergonomics for improved productivity. Springer Nature, pp 869–881

Bhattacharjee S, Santra T, Chatterjee A, Chatterjee S, Banerjee N, Chatterjee S, Mukherjee S (2021) Cognitive status of adult Bengalee males undergoing training in football. In: Ergonomics for design and innovation. Springer Nature, pp 1303–1311

Bianco A, Jemni M, Thomas E, Patti A, Paoli A, Roque JR, Palma A, Mammina C, Tabacchi G (2015) A systematic review to determine reliability and usefulness of the field based test batteries for the assessment of physical fitness in adolescents—The Asso. Project. Int J Occupational Med Environ Health 28:445

Bradley PS, Sheldon W, Wooster B, Olsen P, Boanas P, Krustrup P (2009) High—intensity running in English FA Premier League soccer matches. J Sports Sci 27(2):159–168

Carling C, Orhant E (2010) Variation in body composition in professional soccer players: interseasonal and intraseasonal changes and the effects of exposure time and player position. J Strength Cond Res 24(5):1332–1339

Chaterjee S, Banerjee N, Bhattacharjee S, Santra T, Chatterjee A, Chatterjee S, Saha B, Mukherjee S, Manna I (2014) Sports injuries with special reference to soccer: causes, consequences and prevention strategies. Am J Sports Sci 2(6–1):24–30. ISSN: 2330-8559 (Print); ISSN: 2330-8540 (Online). https://doi.org/10.11648/j.ajss.s.2014020601.15

Chen W, Lin CC, Peng CT, Li CI, Wu HC, Chiang J (2002) Approaching healthy body mass index norms for children and adolescents from health-related physical fitness. Obes Rev 3(3):225–232

Jackson AS, Pollock ML (1978) Generalized equations for predicting body density of men. Br J Nutr 40:497–504

Kale M, Asxcxi A, Bayrak C, Acxikada C (2009) Relationships among jumping performances and sprint parameters during maximum speed phase in sprinters. J Strength Cond Res 23:2272

Krishnan A, Sharma D, Bhatt M, Dixit D, Pradeep P (2017) Comparison between Standing Broad Jump test and Wingate test for assessing lower limb anaerobic power in elite sportsmen. Med J Armed Forces India 73:1 4 0–1 4 5

Kundu S, Banerjee N, Santra T, Chatterjee S, Mukherjee S (2014) Impact of Bharatnatyam dancing exercise on lung function of adult Bengalee females. In: User centred design and occupational wellbeing. McGraw Hill, pp 328–331. ISBN: 978-93-392-1970-3

Masocha V, Katanha A (2014) Anthropometry and somatotype characteristics of male provincial youth league soccer players in Zimbabwe according to playing positions. Int J Sci Res 3:554–547

Mukherjee S, Banerjee N, Chatterjee S, Chatterjee S, Chatterjee A, Santra T, Saha B (2014a) Effect of Bharatnatyam dancing on body composition of Bengalee female children. Am J Sports Sci Med 2:56–59. https://doi.org/10.12691/ajssm-2-1-10

Mukherjee S, Banerjee N, Chatterjee S (2012) Effect of Bharatnattyam dancing on body composition and physical fitness status of adult Bengalee females. Indian J Biol Sci 8:9–15. ISSN: 0972-8503

Mukherjee S, Banerjee N, Chatterjee S, Chatterjee S (2014b) Effect of Kathak dancing on obesity indices in women of sedentary avocations. Sci Culture 80(9–10):279–282. ISSN: 0036-8156

Negra Y, Chaabene H, Hammami M, Hachana Y, Granacher U (n.d.) Effects of high-velocity resistance training on athletic performance in repuberal male soccer athletes. J Strength Conditioning Res 30. https://doi.org/10.1519/JSC.0000000000001433

Raylor K, Donovan M (2013) Offseason fitness training keys to speed and endurance. National Intercollegiate Soccer Officials Association, Akismet: Wordpress

Subramanya NS, Pasodi MS (2011) Training and physical fitness. J Arts Culture 2:43–47 (ISSN: 0976-9862)

Suchy Y, Kraybill ML, Larson JCD (2010) Understanding design fluency: motor and executive contributions. J Int Neuropsychol Soc 16:26–37

Sutton L, Scott M, Wallace J, Reilly T (2009) Body composition of English premier league soccer players: influence of playing position, international status and ethnicity. J Sports Sci 27:1019–1026

Tahara Y, Moji K, Tsunawake N, Fukuda R, Nakayama M, Nakagaichi M, Komine T, Kusano Y, Aoyagi K (2006) Physique, body composition and maximum oxygen consumption of selected soccer players of Kunimi high school Nagasaki, Japan. J Physiol Anthropol 25(4):291–297

Vestberg T, Gustafson R, Maurex L, Ingvar M, Petrovic P (2012) Executive functions predict the success of top-soccer players. PLoS ONE 7(4):1–5

Chapter 2
Doping Awareness, Perception, and Attitude Among Sports Science University Students in Ethiopia

Shumetie Agonafer Babu and Ajita Singh

1 Introduction

The term doping is derived from an ancient African dialect "doop" as meaning a mixture or a potion (Mazzeo et al. 2018). Doping is an ancient phenomenon and athletes sought to improve their own performance by taking mixtures of various types of plant origin (Dimeo and Møller 2018). Doping exists not only in professional sport, but also affects amateur athletes (Mazzeo and Raiola 2018). Doping is thus increasingly a matter that concerns the whole society (Rintaugu and Mwangi 2021). And also use of doping in sports activities has become massive and systematic (Miskulin et al. 2021). Doping possess a threat to sports worldwide. It undermines the principle of open and fair competition (Mwangi et al. 2019). It is a factor that discourages the practice of sport in general and puts the professional under unreasonable pressure and affects individual health (Mwangi et al. 2019). It is a problem of sports organizations (Rintaugu and Mwangi 2021). In order to obtain better results in sports, athletes are seeking it as alternative methods (Dimeo and Møller 2018). Individuals have tried to improve in the artificial way their physical performances (Mazzeo et al. 2018). Reasons are various and changed over time (Dimeo and Møller 2018). Today, doping is the biggest problems in sports (Rintaugu and Mwangi 2021). Even disabled athletes desire to enhance by using doping (Miskulin et al. 2021).

S. Agonafer Babu (✉)
Sport Academy, Department of Sports Science, Debre Markos University, Debre Markos, Ethiopia
e-mail: shumetiea@gmail.com

A. Singh (✉)
Head Department of Sports Science, Faculty of Medicine, Punjabi University, Patiala, India
e-mail: ajitadsingh1@gmail.com

A. D. Singh et al. (eds.), *Sports Ergonomics*, Design Science and Innovation,
https://doi.org/10.1007/978-981-97-7804-1_2

2 Methods and Materials

A cross-sectional study was employed with self-administered and standardized questionnaires 2020/2021 graduate students from 13 governmental universities were invited to participate in the study. These universities were selected using a simple random sampling method. Graduate sports science students were chosen as the study population since they had finalized most of the courses and were likely to become sports science professionals who might be a coach of athletes. A pilot study was conducted. Data were analyzed using SPSS version 24. For all inferential tests, statistical significance was fixed at $p < 0.05$. To examine the association level of students doping APA, chi-squared test was used.

3 Results

As given in Table 1, 58% of females have low doping awareness, whereas male (52.8%) students were reported to have higher doping awareness. This implies that the males have more associated to doping awareness. The statistically test values $x^2 = 1.906$ and p-value (sig.) = 0.016 revealed that significant ($p < 0.05$). Generally, majority of sports science students (55%) had high doping awareness (Fig. 1).

Table 2 revealed that the female (51%) students were reported low doping perception. However, the majority of male (58.5%) students had high doping perception. This indicates that male sports science students have more doping perception. Test of statistical $x^2 = 6.533$ and p-value (sign.) = 0.011 indicated significant ($p < 0.05$). Universal majority of sport science students (54.5%) have better doping perception (Fig. 2).

From Table 3, the female (42.4%) students report a low attitude toward doping. However, male (60.2%) students had great attitude to doping. This designates that male sports science students had an abundant attitude toward doping. Test of statistical $x^2 = 0.483$a and p-value (sign.) = 0.048 revealed significant ($p < 0.05$). More of sports science students (59.1%) scored better doping attitude (Fig. 3).

Table 1 Association of doping awareness among sports science university students in Ethiopia focused on gender

Variable	Gender	Low	High	Total	X^2	P-value
Doping awareness	Male	199 (47.2%)	223 (52.8%)	422 (100%)	1.906	0.016
	Female	182 (58%)	132 (42%)	331 (100%)		
	overall	331 (45%)	405 (55%)	736 (100%)		

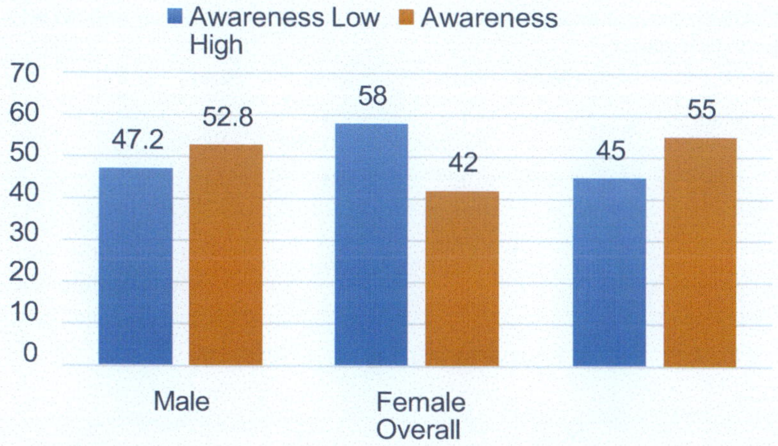

Fig. 1 Graphical representation of sports science university students doping awareness in Ethiopia

Table 2 Association of doping perception among sports science university students between genders in Ethiopia

Variable	Gender	Low	High	Total	X^2	Sign.
Doping perception	Male	175 (41.5%)	247 (58.5%)	422 (100%)	6.533	0.011
	Female	160 (51%)	154 (49%)	314 (100%)		
	Overall	335 (45.5%)	401 (54.5%)	736 (100%)		

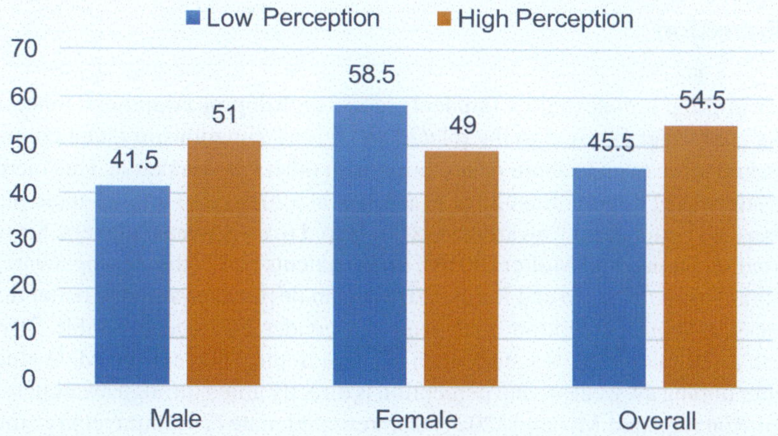

Fig. 2 Graphical representation of the percentage means value of sports science students doping perception in Ethiopia

Table 3 Association of attitude toward doping among sports science university students in Ethiopia in relation to their gender

Variable	Gender	Low	High	Total	X^2	Sign
Attitude toward doping	Male	168 (39.8%)	254 (60.2%)	422 (100%)	0.483	0.048
	Female	133 (42.4%)	181 (57.6%)	314 (100%)		
	Overall	301 (40.9%)	435 (59.1%)	736 (100%)		

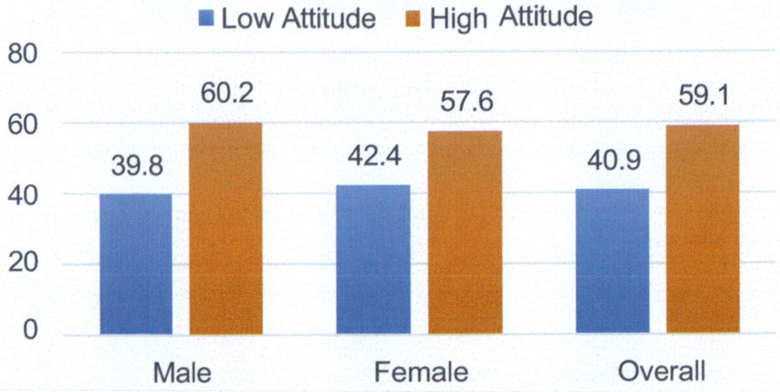

Fig. 3 Graphical representation of the percentage means value of sports science university students' attitude toward doping in Ethiopia

4 Discussion

As given in Table 1, male (52.8%) students scored high doping awareness. Since, male students are further exposed to the related conditions, situations, societal communications, and more open to share information with others as compared. This increases their awareness of doping. Results of Mazzeo et al. (2018) also implies male players had described better doping awareness as compared to their female players. Similarly supported by Dimeo and Møller (2018), male students had better doping awareness. Correspondingly, (Mazzeo and Raiola 2018) the male players validated better doping awareness. Doping perception specifically to gender of students (Table 2) result shows that the male (58.5%) students had more doping perception. Male students had better doping awareness, and perception is directly allied through awareness. The study of Rintaugu and Mwangi (2021) also revealed male (74%) university students had more doping perception. Besides to this (Miskulin et al. 2021), specified male players had better doping perception. Attitude of students to doping (Table 3) indicated that male (60.2%) students had negative feeling to doping. The result might be due to male student sports participation which is good and had better doping awareness and perception; more ideal mentalities toward doping and opportunities. Supported by Mwangi et al. (2019) that male students not permissive attitude to doping agents. In addition (Backhouse and McKenna 2011), investigated male students had better attitude of doping. Similarly, (Brand et al. 2014) result shows male athletes having more negative attitude to PES. In addition the outcomes, (Backhouse et al. 2013; Pöppel and Büsch 2019; Al Ghobain et al. 2016) implies male athletes have a negative attitude toward doping.

5 Conclusion

It could be concluded that the male sports science university students had better association of APA toward doping. Generally, sports science university students had better doping APA. The ETH-NAD and universities gave an attention to female students as well as prepared short- and long-term training, design seminars, and conferences on doping to female students in a special case.

References

Al Ghobain M, Konbaz MS, Almassad A, Alsultan A, Al Shubaili M, AlShabanh O (2016) Prevalence, knowledge and attitude of prohibited substances use (doping) among Saudi sport players. Subst Abuse Treat Prev Policy 11(1):1–6

Backhouse SH, McKenna J (2011) Doping in sport: a review of medical practitioners' knowledge, attitudes and beliefs. Int J Drug Policy 22(3):198–202

Backhouse SH, Whitaker L, Petróczi A (2013) Gateway to doping? Supplement use in the context of preferred competitive situations, doping attitude, beliefs, and norms. Scand J Med Sci Sports 23(2):244–252

Brand R, Wolff W, Thieme D (2014) Using response-time latencies to measure athletes' doping attitudes: the brief implicit attitude test identifies substance abuse in bodybuilders. Subst Abuse Treat Prev Policy 9(1):1–10

Dimeo P, Møller V (2018) The anti- doping crisis in sport: causes, consequences, solutions. Routledge

Mazzeo F, Altavilla G, D'elia F, Raiola G (2018) Development of doping in sports: overview and analysis. J Phys Educ Sport 18(3):1669–1677

Mazzeo F, Raiola G (2018) An investigation of drugs abuse in sport performance

Miskulin I, Grbic DS, Miskulin M (2021) Doping attitudes, beliefs, and practices among young, amateur Croatian athletes. Sports 9(2):25

Mwangi FM, Toriola O, Rintaugu EG (2019) Influence of achievement goals and motivational climate on attitudes toward doping among East African university athletes. Afr J Phys Activity Health Sci (AJPHES) 25(4):547–562

Pöppel K, Büsch D (2019) The doping critical attitude of elite sports coaches in combat sports. Ger J Exerc Sport Res 49(2):168–178

Rintaugu EG, Mwangi FM (2021) Knowledge, attitudes and perceptions on doping among university students in physical education and sport science related degree programmes

Chapter 3
Effect of Herbal Supplements *Prosopis Cineraria* (Khejri) and Spirulina on Muscle Mass, Skeletal Muscle Mass, and Fat-Free Mass in Cricket Players

Anjali Vashishth, Arvind Pareek, Sunil G. Purohit, B. R. Kasvan, and Neha Singh

1 Introduction

Nowadays, herbal dietary supplements are quite famous among athletes. Many reasons behind its fame include increasing energy, mental alertness, promoting muscle growth, inducing weight loss, or instigating many other metabolic and physiological responses which ultimately enhance sports performance. Herbal supplements are derived from the bark, leaves, berries, roots, gums, seeds, stems, or flowers of plants and contain numerous phytochemicals that provide nutritive or medicinal values (Williams 2006; Thakkar et al. 2020; Vinus et al. 2018; Dasgupta 2019; Ng et al. 2021). Herbal or botanical supplements contain vitamins, minerals, amino acids, dietary elements, food components, and proteins, which are used to supplement the diet and performance enhancement and provide some other beneficial effects on our bodies. Herbal supplements are regarded as foods and are assumed to be safe if consumed in a fixed amount (Navarro et al. 2017; Peacock et al. 2019; Karaman et al. 2018; Feng et al. 2021). The use of herbal supplements modulates by Food and Drug Administration (FDA) as a unique category of food and classified it as "dietary supplements" on the report of Dietary Supplement Health and Education Act (DSHEA) of 1994 (Sellami et al. 2018; Balestrino 2019; Calvo 2021; Ibrahim et al. 2021; Denham 2021). *Prosopis cineraria* (Khejri) and Spirulina are also one of famous herbal dietary supplements nowadays and received current attention due

A. Vashishth · S. G. Purohit · B. R. Kasvan · N. Singh (✉)
Department of Sports Bioscience, Central University of Rajasthan, Bandarsindri, Kishangarh, Ajmer, India
e-mail: neha.singh@curaj.ac.in

A. Pareek
Department of Botany, Maharshi Dayanand Saraswati University, Ajmer, India

A. D. Singh et al. (eds.), *Sports Ergonomics*, Design Science and Innovation,
https://doi.org/10.1007/978-981-97-7804-1_3

to their nutrient composition, which has high nutrient content of vitamins, minerals, proteins, essential and non-essential amino acids, phytochemicals, essential fatty acids, fibers, carbohydrate, etc. (Bagheri et al. 2022; Pareek et al. 2015; Yadav et al. 2018).

Prosopis cineraria is the botanical name of the Khejri, a state tree of the Rajasthan and worshipped by many communities. It contains several climatic adaptations, so it can easily survive in wide range of climatic variations. It is also called as "Kalpatru" which denotes "the king of desert" because of its food, feed, and medicinal values. It is cultivated in many countries worldwide, particularly in western and southern Asia including India, Afghanistan,

Iran, Saudi Arabia, Pakistan, etc. (Pathak and Kumar 2017; Baibout et al. 2022; Kumari and Solanki 2021). The crude extracts of *Prosopis cineraria* are an exemplary source of medicine and produce many ergogenic effects in the human body (Girase et al. 2016; Nodushan et al. 2020; Nallasivam and Gokhale 2022; Jain et al. 2020). Ayurveda and other traditional medicinal systems find out the use of herbal supplements to treat various diseases. The bark of Khejri is dry, bitter with a sharp taste, tonic, and helps prevent leprosy, dysentery, asthma, leukoderma, piles, muscle tremors, wandering of mind, etc. The flowers are blended with sugar and used during pregnancy to safeguard against miscarriage. The ashes of its bark rubbed over the skin to remove body hair. The smoke of its leaves is used for the prevention of eye troubles. Every part of Khejri provides positive beneficial effects on the human body (Goyal et al. 2020; Ram et al. 2020; Afifi and Al-rub 2018). Khejri is also used as a dish in Rajasthan, and its pickle is also made (Garg and Mittal 2013; Kaushik et al. 2020; Kumar et al. 2019). A study conducted to see the effect of oral treatment with Prosopis cineraria leaves extract on antidepressant and skeletal muscle relaxant activity in the mice concluded that Prosopis cineraria extract possesses significant antidepressant like effect and skeletal muscle relaxation activity (George et al. 2012). Khejri has lots of beneficial effects on the human body, but in the sports world, Khejri has not undergone any study till now. Many studies proposed the medicinal impact of Khejri, but no studies have been done to evaluate the role of Khejri in the enhancement of sports performance, muscle strength, muscle mass, or any other sports-related parameters.

Spirulina (*Spirulina maxima*) is an alga made up of a precious combination of nutritional substances, including those practically absent in our daily diet. It is the best balanced, highly nutritive herbal dietary supplement that satisfies the demands of all body systems. Energetic value of 10 g dry powder of spirulina is 29 kcal, 0.7 g fat, 5.7 g protein,

2.4 g carbohydrate, 57 IU vitamin A, 0.24 mg vitamin B1, 0.38 mg vitamin B2, 1.3 mg vitamin B3, 1 mg vitamin C, 0.5 mg vitamin E, 136 mg potassium, 12 mg calcium, 20 mg magnesium, 2.8 mg iron, 0.2 mg zinc, etc. It is a highly rich source of β-carotene and iron. The literature data shows that Spirulina improves the condition of the immune system by enhancing white blood cells and red blood cells and assisting detoxification, enhancing the hematopoietic system, increasing resistance to hypoxia, etc. (Milasius et al. 2009; Lafarga et al. 2020, 2021; Grosshagauer et al. 2020; Costa and BCB, Rosa GM, Moraes L, Moraes MG, Mitchell BG. 2019; Jung et al. 2019). A study was performed to see the effects of an exhaustive exercise before and after aerobic training along with dietary Spirulina supplementation on oxidative stress in inactive obese men and concluded that Spirulina supplementation could prevent oxidative damage caused by an exhaustive exercise (Soleymani et al. 2018). A systematic review examined the effect of Spirulina supplementation on (i) exercise performance, (ii) substrate metabolism, and (iii) blood redox status both at rest and after exercise. In the end, they conclude that Spirulina supplementation induced a significant increase in exercise performance, fat oxidation, and growth-stimulating hormone concentration and attenuated the exercise-induced increase in lipid peroxidation (Kalafati et al. 2010). Another study was conducted to see the effects of eight weeks of high-interval intensity training (HIIT) and Spirulina supplementation on immunoglobulin levels, cardio-respiratory fitness, and body composition of overweight and obese women. They conclude that HIIT with Spirulina supplementation affects the body composition but also enhances the immunoglobulin concentration, which regulates the immune system's function (Nobari et al. 2022). Many studies find out the effects of Spirulina supplementation on various sports-related physiological and biochemical parameters, but I did not find any study on cricket players.

Cricket is a famous international team sport with various formats ranging from short duration 20–20 matches to long duration test series matches (Scanlan et al. 2016). Despite the popularity of cricket all over the world and a large number of studies have been done on different fitness parameters of cricket players, but none of the studies investigated the effect of *Prosopis cineraria* and Spirulina supplementation on muscle mass, skeletal muscle mass, and fat-free mass (Pyne et al. 2006). Muscle mass provides strength, energy, and mobility and prevents injuries, which is very important for every player. It also plays a significant role in the high performance of athletes who just need to do proper workouts and design a specific diet plan according to their specific nutrient needs (Amini et al. 2019; Soomro et al. 2018; Constable et al. 2021). The aim of the present study is to investigate the efficacy of *Prosopis cineraria* and Spirulina supplementation on muscle mass, skeletal muscle mass, and fat-free mass. Finally, the objective of the study was to examine the effect of 21 days of *Prosopis cineraria* and Spirulina supplementation on muscle mass, skeletal muscle mass, and fat-free mass in cricket players of Central University of Rajasthan.

2 Methods

2.1 Subjects

Forty university cricket players (age 23.53 ± 0.55 years) participated in this study. Subjects were considered proper cricket players if they had match sessions at least three times a week for at least four years and participation in an intramural cricket league before the study. Inclusion criteria involved subjects not taking any drugs or medications for at least 1 year before enrolment in the study, no known medical health issues or diseases, no smoking or alcohol consumption, and no history of cardio-respiratory disease. Only active cricket players became part of this study. Potential subjects were excluded who failed to meet the inclusion criteria. A written consent form was provided to fill in their general information, and they signed with the condition to use their data for research and agreement for their volunteer participation. The document included an explanation of the risk, benefits of their participation, and the purpose of the study. After consent was obtained, any possible queries were solved to ensure clarification.

2.2 *Study Design*

Figure 1 shows the experimental design of the study. Prior to pre-measurements, subjects were fully explained with all testing and experimental procedures. 40 cricket players were randomly divided into four groups: Group 1 ($n = 10$): Both supplements, Spirulina (A) and Khejri (B), Group 2 ($n = 10$): Supplement Spirulina (A), Group 3 ($n = 10$): Supplement Khejri (B), and Group 4 ($n = 10$): Control. After dividing the subjects, pre-measurements were done to measure muscle mass, skeletal muscle mass, and fat-free mass. Subsequently, subjects began 21 days intervention period in which both supplements were given to individual groups and in combination to see the combined effect. The intervention period was 21 days, and supplements were given in 500 mg doses daily. All subjects had to fill out a 21-day activity diary in which they had to fill in which type of activity they did and for how long by which we determined they were physically active during the intervention period. Post-measurements were done after the completion of the intervention period.

2.3 **Prosopis Cineraria** *and Spirulina Supplementation*

Subjects in Group 1 (Supplement A + B) received one capsule of Spirulina (500 mg) after breakfast and one capsule of Khejri (500 mg) after lunch. Additionally, subjects in Group 2 (Supplement A) received one capsule of Spirulina (500 mg) after breakfast, and subjects in Group 3 (Supplement B) received one capsule of Khejri (500 mg) after breakfast. The control group did not get any supplementation in the intervention period. The selected amount and time of ingestion were based on prior studies on supplements (Koozehchian et al. 2019). Subjects were fed daily by giving water and supplements, but only Group 1 received the supplement Khejri in advance, which they had to take after lunch. The supplementation schedule was also followed up on phone calls and WhatsApp groups to ensure appropriate supervision of the research work.

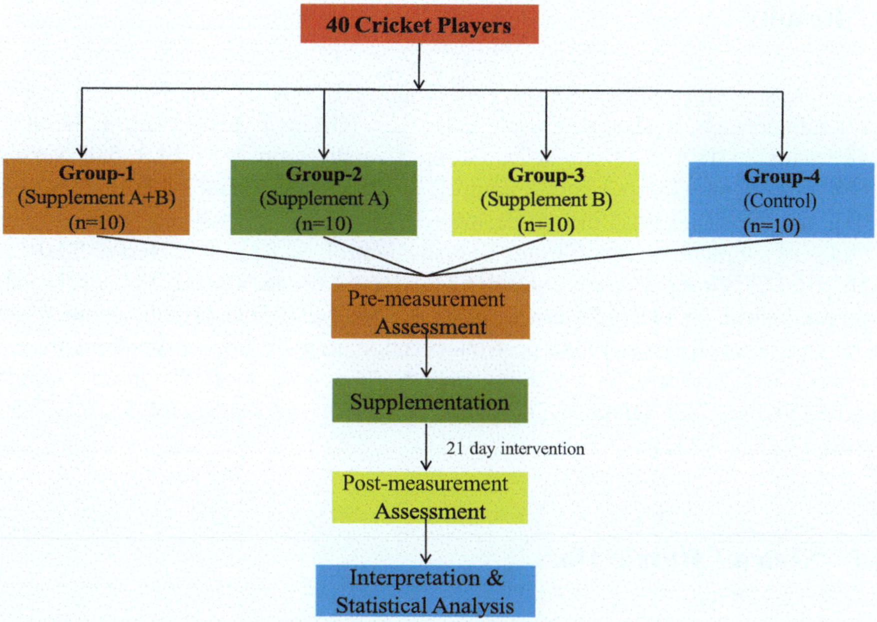

Fig. 1 Schematic of the study design

2.4 *Anthropometric and Body Composition Measurements*

Subjects were instructed to report to the departmental laboratory for testing in a usually hydrated state in the wake of an overnight fast. After the arrival, subjects were mandated to empty their bladder immediately for measurements. The subject's height was measured with the help of a stadiometer nearest to 0.1 cm. Body weight, muscle mass, skeletal muscle mass, and fat-free mass were measured by the remote display version of the Tanita body composition analyzer (DC-430MA III). Measurements were taken in a temperature-controlled laboratory.

2.5 *Statistical Analysis*

Statistical analyses were performed using SPSS software. Shapiro–Wilk test was used to test the normality of data. Descriptive statistics were expressed as mean (±SEM) for each variable. For the data collected in the current study, the mean value, standard deviation, standard error of mean, independent t-test, and one-way ANOVA was applied to analyze the data. Significance was set at an alpha level of 0.05.

3 Results

In comparison with control Group 4, experimental Groups 1, 2, and 3 showed significant enhancement in skeletal muscle mass (Kg) (Group 1 versus control: 34.48 ± 2.48 versus 29.05 ± 2.78, $p < 0.01$; Group 2 versus control: 31.29 ± 5.07 versus 29.05 ± 2.78, ns; Group 3 versus control: 33.42 ± 4.87 versus 29.05 ± 2.78, $p < 0.05$), significant increase in muscle mass (Kg) was found post-intervention after 21 days of supplementation (Group 1 versus control: 54.81 ± 3.97 versus 50.01 ± 4.56, $p < 0.05$; Group 2 versus control: 52.02 ± 6.64 versus 50.01 ± 4.56, ns; Group 3 versus control: 55.14 ± 6.24 versus 50.01 ± 4.56, $p < 0.05$),and significant increase in fat-free mass (Kg) was found post-intervention after 21 days of supplementation (Group 1 versus control: 58.1 ± 3.84 versus 52.76 ± 4.81, $p < 0.05$; Group 2 versus control: 54.14 ± 7.88 versus 52.76 ± 4.81, ns; Group 3 versus control: 57.32 ± 7.59 versus 52.76 ± 4.81, ns).

3.1 *Skeletal Muscle Mass*

As shown in Fig. 2, skeletal muscle mass was found to increase in Group 1 and Group 3 with supplement A and B combination and supplement B alone compared to the control group.

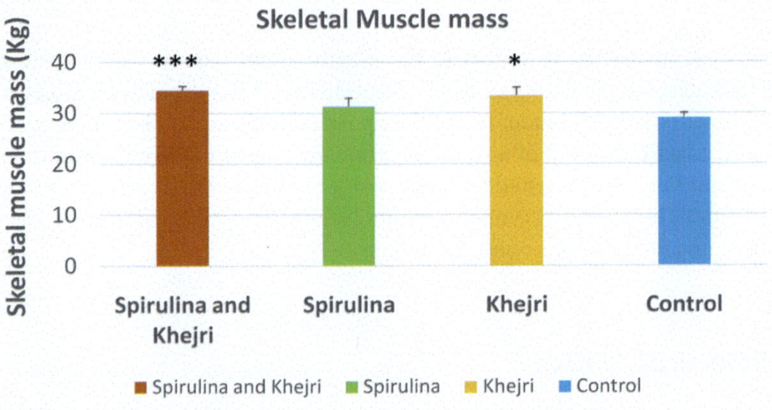

Fig. 2 Skeletal muscle mass comparison between the groups

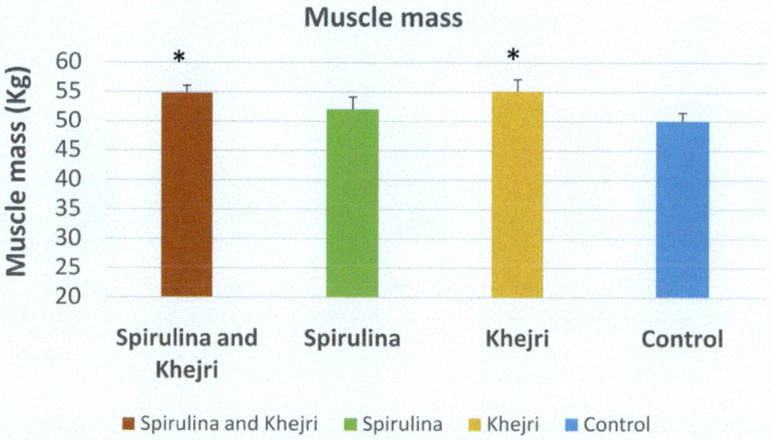

Fig. 3 Muscle mass comparison between the groups

3.2 *Muscle Mass*

As shown in Fig. 3, muscle mass was found to increase in Group 1 and Group 3 with supplement A and B combination and supplement B alone compared to control group.

3.3 *Fat-Free Mass*

As shown in Fig. 4, fat-free mass was found to increase in Group 1 and Group 3 with supplement A and B combination and supplement B alone compared to the control group.

4 Discussion

Athletes go through well-planned regular, intense training, which can quickly lead their body to a state of imbalance and fatigue, affecting their physical abilities and overall performance. Spirulina is a natural herbal supplement that prevents chemical imbalance in the athletic body and contributes to optimal performance in terms of muscle strength. It strengthens the muscles and boosts energy. Khejri (*Prosopis Cineraria*) is a prevalent tree species in India. The pods of Khejri are consumed as vegetables, and leaves are used as traditional medicine. It is observed as one of the valuable plants in the native system of medicine for many areas in the world. Ancient literature on Arabian Gulf and Indian desert exemplifies the importance of Khejri

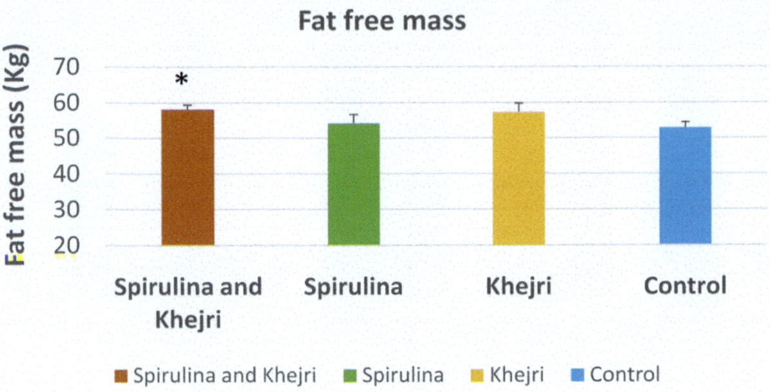

Fig. 4 Fat-free mass comparisons between the groups

in treating various diseases like asthma, dysentery, leukoderma, dyspepsia, leprosy, and earache. In the current study, we explored the ergogenic potential of Spirulina and Khejri as supplements for enhancing skeletal mass, muscle mass, and fat-free mass of cricket players. We also investigated the combined effect of both the above-described supplements. We found that the supplementation of Khejri (500 mg) alone for 21 days was most impactful on the muscle mass, skeletal muscle mass, and fat-free mass of the cricket players. Khejri alone and Khejri in combination with Spirulina were found more effective compared to Spirulina supplementation alone. It could be that if the intervention period of 21 days would have been increased for more days and if the dosage increased to more than 500 mg, Spirulina supplementation alone could have a similar effect as in combination with Khejri. The study indicated the positive effects of these supplements on cricket players. These results and related theories can further be investigated in upcoming studies with more subjects and longer intervention periods.

5 Conclusion

All athletes have different nutritional demands according to the fitness requirements of their respective sports. Based on this study, we can conclude that herbal supplements have beneficial positive effects on cricket players with significant impact on enhancement of muscle mass, skeletal muscle mass, and fat-free mass. A balanced diet with proper nutrients is of utmost importance for players. To achieve their ultimate goal, athletes need to add supplements to their diet to fulfill any lack of nutrients and stay healthy and perform well in regular training sessions and competitions. However, synthetic supplements can produce adverse effects on their health. Our bodies cannot absorb these supplements easily because they have different absorption mechanisms. Therefore, to overcome such limitations of synthetic supplements, herbal supplements are the best option to choose to fulfill dietary requirements which can be quickly absorbed and have negligible side effects. In this study, the ergogenic potential of Spirulina and Khejri supplementation was investigated and established for 21 days of the intervention period. The positive outcome of the above herbal supplements endorses further studies in this direction to promote research on herbal supplements that can be helpful for athletes in improving their health and performance without any side effects.

6 Conflicts of Interest

None.

Acknowledgements The authors thank all the subjects who came forward for participation in the study. We also thank the technical assistance of the staff of the Department of Sports Bioscience, CURAJ, during the study. UGC Startup Grant supported this research.

References

Afifi HSA, Al-rub IA (2018) *Prosopis cineraria* as an unconventional legumes, nutrition and health benefits. Legume Seed Nutraceutical Res. https://doi.org/10.5772/intechopen.79291

Amini B, Boyle SP, Boutin RD, Lenchik L (2019) Approaches to assessment of muscle mass and myosteatosis on computed tomography: a systematic review. J Gerontol Ser 74(10):1671–1678

Bagheri R, Negaresh R, Motevalli MS, Wong A, Ashtary-Larky D, Kargarfard M, Rashidlamir A (2022) Spirulina supplementation during gradual weight loss in competitive wrestlers. Br J Nutr 127(2):248–256

Baibout M, Corcket E, Kothari SL, Fievet V (2022) Ecosystem services provided by *Prosopis cineraria* (L.) Druce in the drylands of Southern and Western Asia. Botany Lett 169(1):30–42

Balestrino M (2019) The risks and uses of dietary supplements. Lancet Neurol 18(9):814–815

Calvo MS (2021) Expanding our understanding of dietary supplement use to include both civilian and institutionalized consumers: the us military dietary supplement use study. J Nutr 151(11):3267–3268

Constable M, Wundersitz D, Bini R, Kingsley M (2021) Quantification of the demands of cricket bowling and the relationship to injury risk: a systematic review. BMC Sports Sci Med Rehabil 13(1):1–12

Costa JAV, Freitas BC, Rosa GM, Moraes L, Moraes MG, Mitchell BG (2019) Operational and economic aspects of Spirulina-based biorefinery. Bioresource Technol 292:121946

Dasgupta A (2019) Antiinflammatory herbal supplements in Translational inflammation. Elsevier, pp 69–91

Denham BE (2021) Dietary supplements in the USA: problematic trends. Public Health Nutr 24(9):2771–2775

Feng Z, Yang J, Xu M, Lin R, Yang H, Lai L, Wang Y (2021) Dietary supplements and herbal medicine for COVID-19: a systematic review of randomized control trials. Clin Nutr ESPEN 44:50–60

Garg A, Mittal SK (2013) Review on *Prosopis cineraria*: a potential herb of Thar desert. Drug Invention Today 5(1):60–65

George M, Joseph L, Sharma A (2012) Antidepressant and skeletal muscle relaxant effects of the aqueous extract of the *Prosopis cineraria*. Braz J Pharm Sci 48:577–581

Girase MV, Jadhav ML, Jain AS (2016) *Prosopis Spicigera*: a nature's gift. Int J Pharm Chem Anal 3(1):49–52

Goyal S, Mukim M, Sharma P, Sharma A, Nagar JC, Jain V, Ahmad N, Singh P (2020) Prosopiscineria: an updated review. Int J Herbal Med 8(1):54–57

Grosshagauer S, Kraemer K, Somoza V (2020) The true value of Spirulina. J Agric Food Chem 68(14):4109–4115

Ibrahim R, Nyska A, Dunnick J, Ramot Y (2021) The toxicologic pathology aspects of selected natural herbal products and related compounds. J Toxicol Pathol 34(3):181–211

Jain R, Jayswal S, Maitreya B (2020) Phytochemical screening of *Prosopis cineraria* L. Druce pods. Int J Res Appl Sci Eng Technol 3(01):70–74

Jung F, Kroger-Genge A, Waldeck P, Kupper JH (2019) Spirulina platensis, a super food? J Cell Biotechnol 5(1):43–54

Kalafati M, Jamurtas AZ, Nikolaidis MG, Paschalis V, Theodorou AA, Sakellariou GK, Koutedakis Y, Kouretas D (2010) Ergogenic and antioxidant effects of spirulina supplementation in humans. Med Sci Sports Exerc 42(1):142–151

Karaman E, Erkin O, Senman S, Yildirim Y (2018) The use of herbal supplements by individuals with diabetes mellitus. JPMA. J Pak Med Assoc 68(4):587–594

Kaushik N, Arya S, Yadav PK, Bharadwaj KK, Gaur RK (2020) Khejri (*Prosopis cineraria* L. Druce) based agroforestry systems in the arid and semi-arid region: supporting ecosystem services. Indian J Agroforestry 23(2)

Koozehchian MS, Kazaj KT, Sarshin A, Gaeini A, Naderi A, Stannard SR, Kaveh M, Chandler TJ (2019) Effects of short-term spirulina supplementation on oxidative stress markers in mountaineers at high altitude. Med Sci Sport Exerc 51:88–88

Kumar M, Govindasamy J, Nyola NK (2019) In-vitro and in-vivo anti-hyperglycemic potential of *Prosopis cineraria* pods extract and fractions. J Biologically Active Products Nat 9(2):135–140

Kumari B, Solanki H (2021) An insight highlight story on the beneficial and phytochemical impact of *Prosopis cineraria* in Tonk, Rajasthan (India). IJBI 3(2)

Lafarga T, Fernandez-Sevilla JM, Gonzalez-Lopez C, Acien Fernandez FG (2020) Spirulina for the food and functional food industries. Food Res Int 137:109356

Lafarga T, Sanchez-Zurano A, Villaro S, Morillas-Espana A, Acien G (2021) Industrial production of spirulina as a protein source for bioactive peptide generation. Trends Food Sci Technol 116:176–185

Milasius K, Malickaite R, Dadeliene R (2009) Effect of spirulina food supplement on blood morphological parameters, biochemical composition and on the immune function of sportsmen. Biol Sport 26(2):157–172

Nallasivam LV, Gokhale JS (2022) Rheological, techno-functional, and physicochemical characterization of *Prosopis cineraria* (Sangri) seed gum: a potential food and pharmaceutical excipient. J Food Process Preserv 46(5):16519

Navarro VJ, Khan I, Bjornsson E, Seeff LB, Serrano J, Hoofnagle JH (2017) Liver injury from herbal and dietary supplements. Hepatology 65(1):363–373

Ng JY, Ahmed S, Zhang CJ (2021) Dietary and herbal supplements for weight loss: assessing the quality of patient information online. Nutr J 20(1):1–13

Nobari H, Gandomani EE, Reisi J, Vahabidelshad R, Suzuki K, Volpe SL (2022) Effects of 8 weeks of high-intensity interval training and spirulina supplementation on immunoglobin levels, cardio-respiratory fitness, and body composition of overweight and obese women. Biology 11(2):196

Nodushan SKH, Emtiazy M, Ardakani MEZ, Salmani MH, Lotfi MH (2020) Comparative effect of *Prosopis cineraria* ointment and eucerin softener on Eczema Area and Severity Index (EASI) score changes in patients with Atopic Dermatitis. Arch Pharm Pract 1:61

Pareek AK, Garg S, Kumar M, Yadav SM (2015) *Prosopis cineraria*: a gift of nature for pharmacy. Int J Pharma Sci Res 6(6):958–964

Pathak V, Kumar P (2017) Phytochemical screening of *Prosopis cineraria* (L.) stem bark and leaves. Int J Innov Sci Res Technol 2(8):306–317

Peacock M, Badea M, Bruno F, Timotijevic L, Laccisaglia M, Hodgkins C, Raats M, Egan B (2019) Herbal supplements in the print media: communicating benefits and risks. BMC Complement Altern Med 19(1):1–14

Pyne DB, Duthie GM, Saunders PU, Peterson CA, Portus MR (2006) Anthropometric and strength correlates of fast bowling speed in junior and senior cricketers. J Strength Conditioning Res 20(3):620–626

Ram H, Jaipal N, Charan J, Kashyap P, Kumar S, Tripathi R, Singh BP (2020) Phytoconstituents of an ethanolic pod extract of *Prosopis cineraria* triggers the inhibition of HMG-CoA reductase and the regression of atherosclerotic plaque in hypercholesterolemic rabbits. Lipids Health Dis 19(1):1–11

Scanlan AT, Berkelmans DM, Vickery WM, Kean CO (2016) A review of the internal and external physiological demands associated with batting in cricket. Int J Sports Physiol Perform 11(8):987–997

Sellami M, Slimeni O, Pokrywka A, Kuvacic G, Hayes LD, Milic M, Podulo J (2018) Herbal medicine for sports: a review. J Int Soc Sports Nutr 15(1):14

Soleymani S, Tofighi A, Babaei Bonab S (2018) Effects of an exhaustive exercise before and after aerobic training along with dietary spirulina supplementation on oxidative stress in inactive obese men. J Appl Health Stud Sport Physiol 5(2):36–44

Soomro N, Strasiotto L, Sawdagar T, Lyle D, Mills D, Ferdinands R, Sanders R (2018) Cricket injury epidemiology in the twenty-first century: what is the burden?. Sports Med 48(10):2301–2316

Thakkar S, Anklam E, Xu A, Ulberth F, Li J, Li B, Hugas M, Sarma N, Crerar S, Swift S, Hakamatsuka T, Curtui V, Yan W, Geng X, Slikker W, Tong W (2020) Regulatory landscape of dietary supplements and herbal medicines from a global perspective. Regul Toxicol Pharmacol 114:104647

Vinus RD, Sheoran N, Mann N, Tewatia B (2018) Potential benefits of herbal supplements in poultry feed: a review. Pharma Innov J 7(6):651–656

Williams M (2006) Dietary supplements and sports performance: herbals. J Int Soc Sports Nutr 3(1):1–6

Yadav E, Singh D, Yadav P, Verma A (2018) Antioxidant and anti-inflammatory properties of *Prosopis cineraria* based phenolic rich ointment in wound healing. Biomed Pharmacother 108:1572–1583

Chapter 4
Kinematic and Neuromuscular Analysis of Lower Limb in Runners with Normal and Excessive Foot Pronation

Rahul Tiwari, Harish Kumar, and Madhvee Tiwari

1 Introduction

Running is popular for its convenience and associated health benefits in a wide range of people. Health benefits of running include improved basal metabolic rate (BMI), endurance, and general fitness/ health (Olin and Gutierrez 2013). However, injuries encountered during running are not uncommon, up to 79% of runners are said to have at least one injury annually (Olin and Gutierrez 2013). Most injuries encountered during running are lower limb injuries to the ankle, knee, or muscular tendinitis or bursitis (McClay and Manal 1998). Several internal and external variables could be to blame for the possibility of injury. Examples of extrinsic risk factors include speed, frequency of training, tools utilized, and breadth of activities, whereas intrinsic risk factors are things like a player's personality and physical characteristics (McClay and Manal 1998; Hintermann and Nigg 1998). Anatomical or biomechanical complications, a lack of flexibility, a lack of strength or muscle imbalance, the type of shoe worn and/or the usage of orthotics, and the type of running terrain are the factors most frequently linked to foot and ankle injuries sustained while running (McClay and Manal 1998, Novacheck 1998). These elements have contributed to the development of injuries, but the quantity of repetitions should also be a consideration (Novacheck 1998). Lower extremity stress fractures are usually linked to improper lower extremity alignment and/or over-pronation (McClay and Manal 1998; Hintermann and Nigg 1998; Novacheck 1998). Overuse injuries of the

R. Tiwari (✉)
Sports Authority of India, NS NIS, Patiala, India
e-mail: rahultiwari.pt@gmail.com

H. Kumar
Department of Sports Science, Punjabi University, Patiala, India

M. Tiwari
MBS Punjab Sports University, Patiala, India

A. D. Singh et al. (eds.), *Sports Ergonomics*, Design Science and Innovation,
https://doi.org/10.1007/978-981-97-7804-1_4

lower extremity are frequently connected with excessive rear-foot pronation (Taunton et al. 2002; Hreljac et al. 2000). The medial muscular structures of the foot and ankle are assumed to be stressed by excessive and/or prolonged pronation, which is also hypothesized to transfer aberrant stresses higher up the kinetic chain. The function of the subtalar joint (STJ), where pronation takes place, has drawn more attention as a result of this (McClay and Manal 1998). Due to the inability to apply external markers to the talus, the STJ is challenging to analyze using conventional motion analysis methods. In order to study the motion of the combined subtalar/talocrural joints, often known as rear-foot motion, researchers have frequently applied markers to the tibia and the calcaneus (McClay and Manal 1998). The majority of the sagittal plane (dorsiflexion/planter-flexion) motion observed between the tibia and calcaneus occurs at the talocrural joint because of the orientation of the axes of the talocrural and STJs. Similar to this, the STJ is where most of the motion in the frontal (inversion/eversion) and transverse (abduction/adduction) planes occurs (McClay and Manal 1998). This has been confirmed by the study of Lundberg, who discovered little relative frontal and transverse plane mobility between the tibia utilizing in vivo techniques (McClay and Manal. 1998). According to sports medicine experts, 26–40% of all running injuries are reportedly due to knee joint dysfunction (McClay and Manal 1998). However, aberrant knee joint mechanics have not received much attention. The sagittal plane is where the majority of knee motion is found. Although this aspect of the motion has been the most carefully examined, the information that has emerged has provided only a limited amount of insight into the causes of injuries connected to running. The talus and, as a result, the tibia rotates inward during closed chain pronation. According to a theory put forth by Engesberg and Andrews (1987), excessive pronation causes excessive internal rotation (IR) of the tibia, which in turn causes aberrant rotational stress to be placed on the musculoskeletal components of the knee joint. More rotation under loaded conditions may result in excessive torsional stresses experienced by the tibia itself. These ideas have yet to be substantiated.

Muscle activation is also one of the important factors, which may lead to the injuries in the runners which is a big concern for athlete and coaches. The biomechanical abnormalities like over-pronation of foot may lead to alteration in the muscular activation of lower extremity (Ahn et al. 2014). Although there are many studies that focus on the kinematics and the neuromuscular activity of the lower extremity in the runners with the pronated foot, none of them have explained both components together (Areblad et al. 1990; Engsberg and Andrews 1987; Ahn et al. 2014). Studies done previously have failed to quantify pronation with motion analysis, as in most 3D studies done on running the foot as has been considered as a single rigid body (Novacheck 1998). Foot models so far have been working with gait analysis with children showing abnormal gaits, and there is a definite dearth of literature on foot mechanics during running. So, the aim of this study is to examine the 2D patterns of motion at the foot and knee and the neuromuscular activation of the lower extremity muscles during running as well as to explore whether differences in these patterns exist between runners with normal rear-foot motion and those who show excessive pronation.

2 Methodology

2.1 Study Design

This study is an experimental design to investigate the kinematics of the ankle joint and knee joint during running from sagittal and frontal plane and muscular activity in hamstring, quadriceps, and gastrocnemius group of muscles.

2.2 Participants

The university ethical approval was taken before recruiting the participants. Total 16 participants including males and females were recruited for the study classifies them into two groups, i.e., with excessive foot pronation (EP) and normal foot pronation (NP) by performing navicular drop test (Magee 2013). The mean age, weight, and height of the participant in NP group ($n = 8$) were 25.63 ± 2.72 years, 72.83 ± 10.79 kgs and 170.83 ± 5.27 cm, respectively, and for group EP ($n = 8$) were 23.25 ± 1.75 years, 76.6 ± 12.79 Kgs, and 164.69 ± 2.84 cm, respectively.

2.3 Instrumentation

Kinematic analysis was done by using quintic motion analysis system (Quintic Consultancy Limited, Sports Motion Capture V21); neuromuscular activity was analyzed by using surface electromyography (MYON AG-m320RX 5vdc/1A/5W Haldenstrasse 5, Switzerland) and was analyzed using ProEMG software. Maximum voluntary isometric contraction (MVIC) for the EMG was measured using an isokinetic dynamometer (Cybex Isokinetic System, Humac Norm Computer System, Serial #- 77161801-2).

2.4 Procedure

Warm-up comprises of cycling for 5 min on a static cycle at 50 RPM to reach about 50–60% of age-related maximum heart rate (Deighan et al. 2012), 10 butt- kicks, 10 high knees, 5 forward lunges, and 5 side lunges on each leg (Green et al. 2011).

Skin preparation: The skin on the thigh and posterior calf was cleaned with alcohol wipes, the hair shaved with razor or cream as comfortable with the participant. This is essential for reducing the skin impedance for application of the surface EMG electrodes (Rainoldi et al. 2004). EMG electrodes were placed on the muscle belly after palpation, approximately 5 cm over the insertion tendon of the muscle or the bulkiest part of the muscle that is being palpated.

Maximum voluntary isometric contraction (MVIC) for static muscle contraction lasting for about 5 s was recorded for each of the above muscle using an isokinetic dynamometer device for comparison with the integral EMG and RMS obtained during running (Deighan et al 2012). The MVIC testing was done for the maximum pronated limb in participants with excessive foot pronation and dominant limb in participants without excessive foot pronation. The surface EMG signals obtained from filtered using a Butterworth low-pass (cut-off 400 Hz, second order) and a Butterworth high-pass filter (cut-off 10 Hz, second order) using the ProEMG software.

Markers were attached to key points on the leg and foot of the participant to help researcher track the movement of the joints and limb during running. The markers was attached to the greater trochanter, lateral knee joint line, lateral malleolus, fifth metatarsal on the training shoe, the lower part of calcaneum/heel, the point of attachment of the tendoachilles to the calcaneum, and posterior aspect of the lower leg on the musculo-tendinous junction of the gastrocnemius. The participant ran on the treadmill at a speed between 6.2 and 7.5 mph (Kranenburg and Smith 1996). 3–5 strides (McClay and Manal 1998) were recorded using quintic cameras from a frontal and a sagittal view; the EMG was recorded for these strides and passed through a high-pass and low-pass filters pipelines (McClay and Manal 1998). The quintic camera recording and EMG recording were time synchronized with a tap at EMG electrode in front of quintic camera.

Data analysis: The video data obtained from the quantic video recording cameras was analyzed using quintic motion analysis software (Quintic Consultancy Limited, Sports Motion Capture v21). Shapiro–Wilk test for normality was performed on all variables, and the data was found not normally distribute; hence, a nonparametric Mann–Whitney two-tailed significance test was performed to verify any differences between the two groups. The statistical significance difference was set at $p \leq 0.05$ for all the tested variables of EMG and 2D motion analysis.

3 Results

3.1 Qualitative Analysis

Entire activity during the single cycle of the running is divided into two main phases (Magee et al. 2010) (Fig. 1), support phase and recovery phase, which are further divided into different subphases—A-Foot strike, B-mid-support, C-Take-off, D-Follow through, E-Forward swing, and F-Foot descent (Fig. 1).

On observation, most of the participant follow the typical running biomechanics with stance and flight phases, except six participant (5 from EP group and 1 from NP group (4 female and 2 males)) who were the front foot runners, putting their front

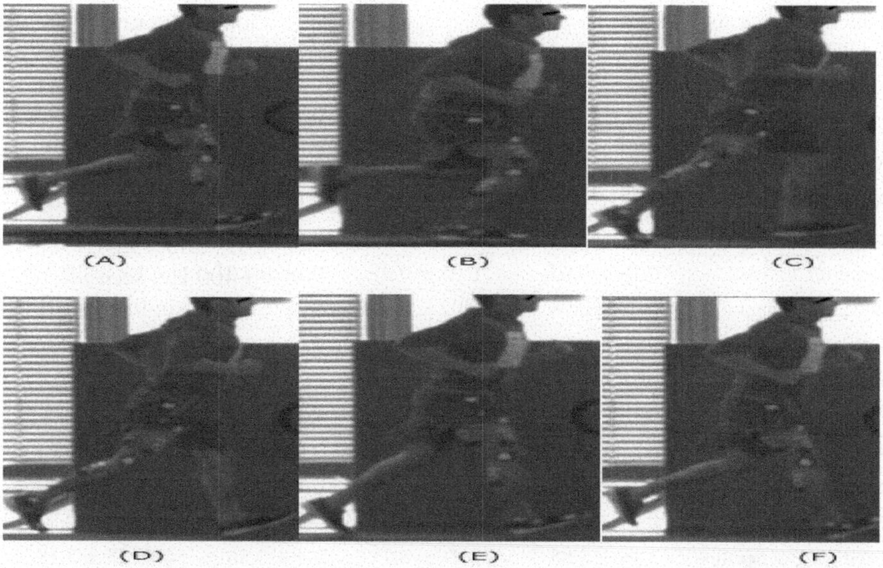

Fig. 1 Different phases of running cycle

foot in contact with ground instead of heel strike at the start of the running cycle. Those six participants were also lacking behind in the time period of total airborne phase, as compared to other 10 participants who were following the proper sequence of the running biomechanics. It is also observed that the participants with the front foot running runs at slower pace than the participant who were following the typical phases of the running.

3.2 Quantitative Analysis EMG Activity

Table 1 presents the maximum voluntric isometric contraction (MVIC) activities during running of the selected muscle with respect to integral and RMS, and there is no significant difference found between the group although there is notable difference found in vastusmedialis, semi-tendinosis, and medial head of gastrocnemius between the group which represents the lower strength of the muscle in the EP group as compare to NP groups. Table also shows the EMG activity of the selected muscle during three running consecutive cycle of the runner of both NP and EP group. None of the muscle is showing the significant difference between the groups although vastusmedialis, semi-membranous, and semi-tendinosis show the lower value of integral by time in EP group while RMS by time is more or less same in the both groups. The difference in the above mention muscle shown in the group is susceptible to injury, and these lower values put extra workload on different muscle groups.

Table 1 EMG activity of different muscles

MVIC				
Muscles	NP group		EP group	
	Integral/MVIC	RMS/MVIC	Integral/MVIC	RMS/MVIC
Vastus medialis	0.19 ± 0.04	0.05 ± 0.01	0.15 ± 0.04	0.04 ± 0.01
Vastus lateralis	0.17 ± 0.03	0.04 ± 0.01	0.17 ± 0.02	0.04 ± 0.01
Semi-membranous	0.22 ± 0.02	0.06 ± 0.01	0.17 ± 0.05	0.05 ± 0.01
Semi-tendinosis	0.18 ± 0.03	0.05 ± 0.01	0.14 ± 0.04	1.8 ± 2.5
Med. gastrocnemius	0.08 ± 0.08	0.02 ± 0.01	0.07 ± 0.03	0.02 ± 0.01
Lat. gastrocnemius	0.06 ± 0.07	0.02 ± 0.01	0.04 ± 0.02	0.01 ± 0.01
Running				
Vastus medialis	0.78 ± 0.9	3.47 ± 4.98	0.57 ± 0.27	1.37 ± 0.46
Vastus lateralis	0.82 ± 0.89	1.78 ± 1.76	0.55 ± 0.24	1.32 ± 0.55
Semi-membranous	0.71 ± 0.44	2.62 ± 3.19	0.96 ± 0.49	2.58 ± 1.69
Semi-tendinosis	0.88 ± 0.39	1.72 ± 0.84	0.67 ± 0.31	0.88 ± 0.87
Med. gastrocnemius	3.45 ± 3.85	8.21 ± 8.75	2.4 ± 2.86	6.25 ± 7.18
Lat. gastrocnemius	3.79 ± 3.73	7.57 ± 7.74	3.33 ± 2.79	8.03 ± 6.22

3.3 Kinematic Analysis

Table 2 presents the average joint angle and linear kinematic during running. There is a significant difference found in ankle joint and tibia-calcaneal range of motion but not in knee joint range of motion in EP group. There is significant difference found in linear acceleration during running, but there is no significant difference found in linear displacement and linear velocity. Both the group shows same linear displacement while running but the EP group is showing slightly higher linear velocity.

Table 2 Kinematic analysis during running

		NP group (mean values)	EP group (mean values)
Joint angle	Ankle	49.32 ± 13.63	69.43 ± 5.01**
	Knee	72.59 ± 13.17	69.43 ± 12.23
	Tibia-calcaneal	21.57 ± 4.81	27.12 ± 6.69**
Linear analysis	Displacement	3.59 ± 0.56	3.59 ± 0.73
	Velocity	3.20 ± 0.43	3.54 ± 0.96
	Acceleration	34.08 ± 5.62	43.49 ± 13.05**

Significant differences are marked by ** ($p < 0.05$)

4 Discussion

The general kinematic analysis of the motion of the rear-foot noted in the present study was consistent at some point and opposite with response to some variables with the few studies in the literature addressing running with the excessive foot pronation and with the normal foot pronation (McClay and Manal 1998; Areblad et al. 1990; Clarke et al. 1984, Engsberg and Andrews 1987; Deighan et al. 2012; Soutas-Little et al. 1987). STJ pronation, which is the combined motion of dorsiflexion, eversion, and abduction (Hertel 2002), occurs during the first half of the stance. As stated earlier, difficulty in placing an external marker on the talus limits our ability to measure true subtalar (talo-calcaneal) motion. However, motion of the rear-foot (tibio-calcaneal joint) appeared to follow the same pattern as the STJ with the motion of pronation (dorsiflexion, eversion, and abduction) occurring in the first half of stance followed by the components of supination (planter-flexion, inversion, and adduction) in the latter half.

4.1 Joint Movement During Running

The amount of joint movement that occurs at the hip, knee, and ankle during the running cycle is different from that which occurs during walking cycle (Magee et al. 2010). At the instant of foot strike while running, the knee joint is already in at least 20° of flexion. This flexed position of the knee is important to assist the runner in attenuating impact forces that are being applied through the lower extremity. Unfortunately, this greater degree of knee flexion also causes increased compressive forces between the patella and femur. This is an important factor, considering the extremely high incidence of knee injuries in runners, particularly to the patella-femoral joint. This flexed position of the knee joint at foot strike also implies that the flexibility of the soleus muscle is just as important as the flexibility of the gastrocnemius when assessing range of motion in runners.

To effectively manage overuse injuries associated with running, it is important to have an understanding of the changes in range of motion at the hip, knee, and ankle joints throughout the different phases of the gait (Deighan et al. 2012). From foot strike to mid-support, hip joint flexion decreases from 45 to 20°, whereas knee joint flexion increases from 20 to 40°. The position of the ankle (talocrural) joint at foot strike is 5° planter-flexed, with progression to 10° dorsi-flexion by the mid-support phase. From mid-support to takeoff, the hip joint moves from 20° of flexion to 5° extended position, helping to propel the runner's mass forwarded in preparation for the next foot strike. The knee joint decreases the amount of its flexion by moving from 40° to a 15° position. Finally, the talocrural joint moves from a 10° dorsi-flexed position to 20° dorsi-flexed position (Deighan et al. 2012; Magee et al. 2010).

During the flight or recovery phase, the swinging limb assists the forward momentum of the runner, thereby improving efficiency. During the follow-through phase of recovery, the hip joint, starting from a 5° position, moves into further hyperextension of 20° supporting the importance of proper gluteus maximus strength and neuromuscular recruitment (Cote et al. 2005). This amount of knee joint motion is significantly greater than that during walking, which is only 65° to 70° and is related to the greater amount of momentum generated by lower limb during running. Finally, the talocrural joint will move from 30° of planter-flexion to a position of neutral (0°). Finally, foot descent, the hip joint will move from a 65° flexed position to 40° of flexion, whereas the knee joint moves from 130° of flexion to 20° of flexion in preparation for foot strike. The talocrural joint will move from 0° to 5° of dorsiflexion to 5° of planter-flexion in preparation for foot strike (Cote et al. 2005).

4.2 *Foot Pronation During Running*

The rate or speed and amount of pronation increase during running. Furthermore, the concept of re-supination after the mid-support phase of the stance cycle is an important consideration in two-dimensional gait analysis (McClay and Manal 1998). As previously stated, no published studies to date on this concept, but the assumption is that the delay in the external rotation of the tibia has the potential for increasing soft tissue stresses on proximal structures (Cote et al. 2005).

Another foot evaluation concept to consider is the onset at which the majority of subtalar joint pronation occurs. Whether the majority of pronation occurs early in the stance phase (early pronation) or late in the stance phase (late pronation) may dictate the need for a footwear modification or the addition of a customized foot orthosis (Magee et al. 2010; Cote et al. 2005; McClay and Manal 1998). Early pronation is the amount of movement measured between foot strike and mid-support of stance phase. Late pronation is the amount of movement measured after mid-support leading up to takeoff.

4.3 *Muscle Activity During Running*

In general, investigations of muscle action during running have demonstrated increased magnitude of EMG activity in comparison with walking (Murley et al. 2009; Hintermann and Nigg 1998). In addition, the duration of EMG activity has been found to be longer during the support phase of running in comparison with the stance phase of walking, and also there are variations in comparison between the runner with normal and excessive foot pronation. However, majority of researcher have used surface electrodes, rather than indwelling or fine electrodes, to study EMG activity because of the velocity and intensity of the motions required for running.

Around knee joint, at the time of foot strike, the quadriceps muscle is very active in preparation for the rapid loading that occurs at this time. The quadriceps muscle group reaches peak activity during the early support phase. Quadriceps muscle activity during this period of running cycle facilitates the shock attenuation process as the hip, knee, and ankle joints undergo flexion to create a shortening of the lower extremity. The hamstring muscle have been shown to be active from before foot strike until the hip joint completes its period of extension later in the support phase.

At ankle joint, the phasic activity of muscles during running is very similar to that during walking, except that the magnitude of activity is increased during running. As the speed of running increases, so does the integrated EMG activity. McCly et al. in 1998 showed that integrated EMG activity increased during the swing phase but remain constant during the contact phase as running speed increased from 3.7 to 9.3 m/s. (Murley et al. 2009; Hintermann and Nigg 1998).

Rear-foot differences between the two groups were most notable in the frontal plane motion, as evidenced by the 1.5-fold increases in the peak eversion noted in the EP group ($n = 8$). These subjects, therefore, landed and pushed off in an everted position. It has been suggested that, when the rear-foot is everted, the arch is lowered and the foot becomes a mobile adaptor that allows for shock attenuated and accommodation to uneven surfaces (Hertel 2002; McClay and Manal 1998). During rear-foot inversion, the arch is raised and the bones of the mid-foot becomes 'locked up', hence allowing the foot to become more stable. This relationship between rear-foot and mid-tarsal joint motion was demonstrated, in vivo, by Lundberg in 1989.

The EMG activity of the selected muscles does not show any significant difference between the EP and NP group, although there is a little difference found in the integral by time in vastusmedialis, semi-membranous, and semi-tendinosis. The above result of the EMG activities of the following muscles are opposite than the result which was found by McClay et al. in 1998, showing significant difference in most of the muscles during running with excessive foot pronation, suggesting more activities of the selected muscle leading toward injury.

4.4 Limitations of the Study

There are several limitations of the present study which need to be considered. The result of the study was based on small sample size, so the findings cannot be generalized and care should be taken before drawing any concrete decision from the study. Hip kinematics is not considered in the present study, which can be the scope for further study. Thus, outcome of the study should motivate future investigations with higher sample size and can be more specific to come to a conclusion, which can be applied to the clinical set up.

5 Conclusion

The present study concludes that there is not at all significant difference found in the EMG activities of the selected muscles but found some in the other parameters like ankle joint ROM, linear acceleration, and pronation angle at mid-stance. The findings of the study suggest that the foot alignment and technique play an important role in the running.

References

Ahn AN, Brayton C, Bhatia T, Martin P (2014) Muscle activity and kinematics of forefoot and rearfoot strike runners. J Sport Health Sci 3(2):102–112

Allinger TL, Engsberg JR (1993) A method to determine the range of motion of the ankle joint complex, in vivo. J Biomech 26(1):69–76

Areblad M, Nigg BM, Ekstrand J, Olsson KO, Ekström H (1990) Three-dimensional measurement of rearfoot motion during running. J Biomech 23(9):933–940

Cavanagh PR, Williams KR (1982) The effect of stride length variation on oxygen uptake during distance running. Med Sci Sports Exerc 14(1):30–35

Clarke T, Frederick E, Hamill C (1984) The study of rearfoot movement in running, in "sport shoes and playing surface". In: Frederic E (ed) Human kinetics, Illinois, p 166

Cole GK, Nigg BM, Ronsky JL, Yeadon MR (1993) Application of the joint coordinate system to three-dimensional joint attitude and movement representation: a standardization proposal

Cote KP, Brunet ME, Gansneder BM, Shultz SJ (2005) Effects of pronated and supinated foot postures on static and dynamic postural stability. J Athl Train 40(1):41

CYBEX ISOKINETIC SYSTEM (2014) Humac norm: Testing and rehabilitation system. [online] Cybex Isokinetic System. Available from: http://www.csmisolutions.com/sites/default/files/humac_norm_brochure_0.pdf

Deighan MA, Serpell BG, Bitcon MJ, Croix MDS (2012) Knee joint strength ratios and effects of hip position in rugby players. J Strength Conditioning Res 26(7):1959–1966

Engsberg JR, Andrews JG (1987) Kinematic analysis of the talocalcaneal/talocrural joint during running support. Med Sci Sports Exerc 19(3):275–284

Evans AM, Copper AW, Scharfbillig RW, Scutter SD, Williams MT (2003) Reliability of the foot posture index and traditional measures of foot position. J Am Podiatr Med Assoc 93(3):203–213

Green BS, Blake C, Caulfield BM (2011) A comparison of cutting technique performance in rugby union players. J Strength Conditioning Res 25(10):2668–2680

Hertel J (2002) Functional anatomy, patho-mechanics, and pathophysiology of lateral ankle instability. J Athl Train 37(4):364

High DM, Howley ET, Franks BD (1989) The effects of static stretch- ing and warm-up on prevention of delayed-onset muscle soreness. Res Q Exerc Sport 60(4):357–361

Hintermann B, Nigg BM (1993) Ponation aus der Sicht der Bewegung sübertragung zwischen Kalkaneus und Tibia. Schweizerische Zeitschrift Für Sportmedizin 41(4):151–156

Hintermann B, Nigg BM, Sommer C, Cole GK (1994) Transfer of movement between calcaneus and tibia in vitro. Clin Biomech 9(6):349–355

Hintermann B, Nigg BM (1998) Pronation in runners. Sports Med 26(3):169–176

Hreljac ALAN, Marshall RN, Hume PA (2000) Evaluation of lower extremity overuse injury potential in runners. Med Sci Sports Exerc 32(9):1635–1641

Karatsolis K, Nikolopoulos CS, Papadopoulos ES, Vagenas G, Terzis E, Athanasopoulos S (2009) Eversion and inversion muscle group peak torque in hyperpronated and normal individuals. Foot 19(1):29–35

Kosterina N, Wang R, Eriksson A, Gutierrez-Farewik EM (2013) Force enhancement and force depression in a modified muscle model used for muscle activation prediction. J Electromyogr Kinesiol 23(4):759–765
Kranenburg KJ, Smith DJ (1996) Comparison of critical speed deter- mined from track running and treadmill tests in elite runners. Med Sci Sports Exerc 28(5):614–618
Lundberg A (1989) Kinematics of the ankle and foot: in vivo roentgen stereo- photogrammetry. Acta Orthop Scand 60(sup233):1–26
Magee DJ (2013) Orthopedic physical assessment. Elsevier Health Sciences
Magee DJ, Zachazewski JE, Quillen WS, Manske RC (2010) Athletic and sport issues in musculoskeletal rehabilitation. Elsevier Health Sciences
MAYO CLINIC STAFF (2014) Electromyography. [Online] Mayo Foundation for Medical Education and Research. Available from: http://www.mayoclinic.org/testsprocedures/electroconvulsive-therapy/basics/definition/prc-20014183
McClay I, Manal K (1998) A comparison of three-dimensional lower extremity kinematics during running between excessive pronators and normals. Clin Biomech 13(3):195–203
Menz HB, Munteanu SE (2005) Validity of 3 clinical techniques for the measurement of static foot posture in older people. J Orthop Sports Phys Ther 35(8):479–486
Murley GS, Landorf KB, Menz HB, Bird AR (2009) Effect of foot posture, foot orthoses and footwear on lower limb muscle activity during walk- ing and running: a systematic review. Gait Posture 29(2):172–187
Nigg BM (2001) The role of impact forces and foot pronation: a new paradigm. Clin J Sport Med 11(1):2–9
Nigg BM, Cole GK, Nachbauer W (1993) Effects of arch height of the foot on angular motion of the lower extremities in running. J Biomech 26(8):909–916
Novacheck TF (1998) The biomechanics of running. Gait Posture 7(1):77–95
Olerud CLAES, Rosendahl YLVA (1987) Torsion-transmitting properties of the hind foot. Clin Orthop Relat Res 214:285–294
Olin ED, Gutierrez GM (2013) EMG and tibial shock upon the first at- tempt at barefoot running. Hum Mov Sci 32(2):343–352
QUINTIC (2014) Quintic USB2 live camera tutorial, sports motion capture V21. [Online] Quintic Consultancy Limited. Available from: http://www.quintic.com/downloads/Quintic%20High-Speed%20Camera/Quintic%20High-Speed%20Live%20 Camera%20Tutorial.pdf
Psycharakis S (2014) Dynamics of vertical jumps. Quintic, Edinburgh
Rainoldi A, Melchiorri G, Caruso I (2004) A method for positioning electrodes during surface EMG recordings in lower limb muscles. J Neurosci Methods 134(1):37–43
Soutas-Little RW, Beavis GC, Verstraete MC, Markus TL (1987) Analysis of foot motion during running using a joint co-ordinate system. Med Sci Sports Exerc 19(3):285–293
Taunton JE, Ryan MB, Clement DB, McKenzie DC, Lloyd-Smith DR, Zumbo BD (2002) A retrospective EP-NP analysis of 2002 running injuries. Br J Sports Med 36(2):95–101
Vinicombe A, Raspovic A, Menz HB (2001) Reliability of navicular displacement measurement as a clinical indicator of foot posture. J Am Podiatr Med Assoc 91(5):262–268

Chapter 5
The Effects of Muscle Energy Technique on Range of Motion and Pain in Patients with Post Arthroscopic Knee Surgery Stiffness

Deepak Kumar Mallick, Sohini Paul, and Tirthankar Ghosh

Muscle Energy Technique (MET) is a form of manual therapy that uses muscles' own energy in the form of gentle isometric contraction to relax the muscle via reciprocal or autogenic inhibition and lengthen the muscle. Post-operatively the complications of knee arthroscopy are nerve damage, decreases the range of motion, persistent pain, and stiffness of the joint. This study aims at the effects of MET on pain and Range of motion given in early rehabilitation.

In the Current Study Twenty-five patients who had prevalence musculoskeletal complications after knee arthroscopy surgery was selected. Out of which eighteen are male and seven are female. To identify the prevalence each patient's range of motion and VAS Score for pain sensation were recorded. The same patients have introduced to MET for another seven days. A statistical test was performed to identify the significant difference between pre and post-values.

Most of the patients were having an average Range of motion (Flexion) (120 ± 9.02), Range of motion (Extension) of 78.00 ± 13.61, and VAS score was (4.32 ± 1.02). Most patients were significant improvement in Range of motion and a decrease in pain sensation as per the VAS Score.

This study shows MET is effective in improving range of motion and Pain in Post-surgical knee stiffness patients in the early stage of rehabilitation.

D. K. Mallick (✉) · S. Paul · T. Ghosh
Ergonomics and Exercise Physiology Laboratory, Faculty of Health and Wellness, Sri Sri University, GodiSahi, Cuttack, Odisha 754006, India
e-mail: mallickdeepak58@gmail.com

T. Ghosh
e-mail: tirthag@srisriuniversity.edu.in

A. D. Singh et al. (eds.), *Sports Ergonomics*, Design Science and Innovation,
https://doi.org/10.1007/978-981-97-7804-1_5

1 Introduction

The knee joint is one of the largest and most complex joints in the body. The knee joins the thigh bone (femur) to the shin bone (tibia). The ligaments join the knee bones and provide stability to the knee such as the anterior cruciate ligament, posterior cruciate ligament, medial collateral ligament, and lateral collateral ligament Matthew Hoffman (2021). The common injury in the knee joint is ACL injury which is followed by arthroscopic surgery for reconstruction of the ACL tear. The ligament tear mainly occurs due to road traffic accidents, or playing games with an incidence of 3–5% in the general population.

Knee arthroscopy is a surgical technique that can diagnose and treat problems in the knee joint. Knee arthroscopy is a surgical procedure used to visualize, diagnose, and treat various problems inside the knee joint. Arthroscopy can be done without fear of infection or of trauma to the joint. Arthroscopic knee surgery is used to treat common knee disorders including meniscus tear and patella problems (Burman et al. 1934). Arthroscopy is a technique of surgery on the joints in which the tip of a thin (4 mm diameter) telescope called an arthroscope which is introduced in to joint and inside the joint examined Maheswari and Mahaskar (2015). One must feel some kind of pain and discomfort post-operatively. The post-operative complications are nerve damage, which decreases range of motion (ROM), persistent pain, and stiffness of joints. By applying appropriate treatment, a physical therapist can do much to enhance the speed and degree of recovery from stiffness. Normal ROM of knee flexion is 90°–13°; knee extension is 135°–0° Norkin and White (n.d.).

Muscle energy technique (MET) is a manipulative procedure that has evolved out of osteopathic medicine and is designed to lengthen muscle and fascia and mobilize joints (Kisner and Colby 2012). MET can help to release and relax the muscle and improve joint range of motion (JROM). One of the main uses of this method is to normalize JROM further than increase flexibility, and techniques can be used on any joints with restricted ROM identified during the passive assessment (Brontzman and Wilk 2003).

This study is focused to see the effects of early intervention of MET and comparing the effects with MET given one week later in the stage of rehabilitation and pain, ROM, and function in patients with post-surgical knee stiffness. MET involves the subject voluntarily contracting the muscle in a precisely controlled direction against the therapist's counterforce (Faqin et al. 2019). Its therapeutic effects are to reduce pain, reduce muscle tone, stretch and tighten muscles, strengthen the weak muscles, improve local circulation, and mobilize joint restriction (Fryer and Ruszkowski 2004).

Relaxation of the antagonist's muscle occurs due to actively contracting the agonist's muscle. This facilitates mobility at the joint due to reciprocal inhibition (Ylinen 2008). There are different interventions that are practiced for the management of knee stiffness which include therapeutic exercise, stretching, strengthening, CPM, and other electrotherapy modalities (Mac Dermid et al. 2012).

2 Methods

A randomized trial was done over a period of three months on subjects aged between 18 and 40 years who were undergone surgery included in this random sample which was taken. The randomization was done by the primary investigator prior to the minimal assessment.

2.1 Outcome Measures

The visual analog scale (VAS) was used to assess the change in pain intensity knee joint range of motion which was measured using the universal half goniometer (Prince 1983). The VAS is a reliable and good tool and is widely used in clinical to measure the intensity of pain. It also helps researchers measure acute and chronic pain (Carlsson 1983; Breivik et al. 2008). Measuring the JROM at the knee using a universal goniometer is a simple and reliable clinical tool (Chapleau et al. 2011).

2.2 Sample Size and Techniques

Twenty-five patients who had prevalence of musculoskeletal complication after knee arthroscopy was selected. Out of which eighteen are male and seven are female.

MET was given by a trained physiotherapist who has knowledge about post-isometric relaxation for seven days with a 5–7 s hold for 8–10 repetitions followed by passive stretching after post-immobilization (Chaitow 2013).

The therapeutic exercise program was given to subjects for at least seven days as follows: Active and passive flexion in supine lying.

Active ankle–foot-toe movement while the subject is in a sitting position. Forced passive movement with a bearable range in prone lying.

All the exercises should be done about ten repetitions each. The pre- and post-exercise assessments were done for seven days. Measurement was taken with the help of another trained physiotherapist who has knowledge about pre- and post-interventions.

3 Statistical Analysis

Data were analyzed using IBM SPSS Statistics for Windows, version 25. For pain, the VAS scale was done, and ROM with the collected data was done using a t-test.

Rom (Extension)

Test Value = 25

t		df	Sig. (2-tailed)	Mean Difference	95% Confidence interval of the difference	
					Lower	Upper
ROM PRE-TEST	40.598	24	0.000*	76.60000	72.7058	80.4942
ROM POST-TEST	19.461	24	0.000*	53.00000	47.3793	58.6207

* = 0.05. The result is significant at $p < 0.05$

Rom (Flexion)
Test Value = 25

t		df	Sig. (2-tailed)	Mean Difference	95% Confidence interval of the difference	
					Lower	Upper
ROM PRE-TEST	40.598	24	0.000*	76.60000	72.7058	80.4942
ROM POST-TEST	48.653	24	0.000*	87.80000	84.0754	91.5246

* = 0.05. The result is significant at $p < 0.05$

Vas Scale
Test Value = 25

t		df	Sig. (2-tailed)	Mean difference	95% Confidence interval of the difference	
					Lower	Upper
VAS PRE-TEST	– 71.663	24	0.000*	– 18.28000	– 18.8065	– 17.7535
VAS POST-TEST	– 100.431	24	0.000*	– 20.68000	– 21.1050	– 20.2550

* = 0.05. The result is significant at $p < 0.05$

4 Results

Table 1 represents the evaluation of pain using the VAS scale.

Table 2 represents the evaluation of ROM (Flexion) by using a goniometer.

Table 3 represents the evaluation of ROM (Extension) by using a goniometer.

Table 1 PRE and POST VAS scores

N		Minimum	Maximum	Mean	Std. deviation
VAS PRE-TEST	25	5.00	9.00	6.7200	1.27541
VAS POST-TEST	25	3.00	6.00	4.3200	1.02956

Table 2 PRE and POST score of ROM flexion

N		Minimum	Maximum	Mean	Std. deviation
ROM PRE-TEST	25	90.00	120.00	101.6000	9.43398
ROM POST-TEST	25	100.00	130.00	112.8000	9.02312

Table 3 PRE and POST score of ROM extension

N		Minimum	Maximum	Mean	Std. deviation
ROM PRE-TEST	25	90.00	120.00	101.6000	9.43398
ROM POST-TEST	25	60.00	110.00	78.0000	13.61678

5 Discussion

The present-day study was undertaken to study the effects of MET when applied immediately after one week of post-operative immobilization and after one week of rehabilitation on some parameters such as pain, ROM of the knee, and muscle testing. MET is shown to be effective in various stages of rehabilitation. In this study, during the early intervention phase, there was no adverse action such as increasing pain, irritation, difficulty, and discomfort.

The study concluded that MET was more effective in improving the stiffness, pain, and ROM, of those who had restricted knee JROM after knee arthroscopy was done in primary stage of ligament tear and any injury to the knee joint in acute condition.

5.1 Effect of Immobilization

After surgery, the patient was kept in immobilization to prevent further damage. But due to long-term immobilization, the protean changes that saliently occur are fibro fatty proliferation of the connective tissue in the joint space and its adherence to the cartilage surface, synovial fluid adhesions, cartilage atrophy, cellular, and fibrillar ligament disorganization, osteoclastic resorption of bone leading to weakling of the ligament insertion (Akeson et al. 1987).

5.2 Reduction in Pain

The intensity of pain was reduced significantly in post-rehabilitation with compare to pre-rehabilitation program. The mean difference is 6.72 ± 1.27 in pre-test evaluation to mean difference is 4.32 ± 1.02 in post-test evaluation with $p < 0.05$.

On application of MET, there may be a reduction in pro-inflammatory cytokines and it may also desensitize the peripheral nociceptors. Blood and lymphatic flow rates may also be affected due to rhythmic muscle contraction, and there could be changes in the interstitial pressure and an increase in the transcapillary blood flow (Fryer 2011). When isometric contraction and stretching were given simultaneously, the muscle and joint receptors are stimulated more and more than with their elasticity zone.

5.3 Improvement of Joint Range of Motion

This study shows that there was an improvement in knee JROM in both flexion and extension after giving therapeutic exercise. The mean value of flexion has been changed from 101.600 ± 9.43 to 120 ± 9.02 with $p < 0.05$, whereas the mean value of extension has been changed from 101.600 ± 9.43 to 78.00 ± 13.61 with $p < 0.05$.

Through deep tissue stimulation and localized vasodilation, MET suppresses sympathetic tone (Schleip 2003). Active muscle contraction has been shown to have neurophysiological effects, including pain inhibition, thus following the muscle to be stretched further (Carlsson 1983; Phadke et al. 2016).

6 Conclusion

There was an improvement in pain, JROM, and function at the knee joint when MET was started immediately after immobilization removal, and later a significant improvement was shown after the MET was given for seven days. MET can be used as an adjacent to the rehabilitation protocol to treat knee arthroscopy patients with stiffness and can be safely given in the early rehabilitation phase of knee pain and decrease ROM and stiffness.

References

Akeson WH, Amiel D, Abel MF, Garfin SR, Woo SL (1987) Effects of immobilization on joints. Clin Orthop Relat Res 219:28–37

Breivik H, Borchgrevink PC, Allen SM et al (2008) Assessment of pain. Br J Anaesth 101(1):17–24

Brontzman S, Wilk K (2003) Clinical orthopedics rehabilitation, pp 227–231

Burman MS, Finkelstein H, Mayer L (1934) J Bone Joint Surg P25-268 16(2)

Carlsson AM (1983) Assessment of pain. I. Aspects of the reliability and validity of the visual analog scale. Pain 16:87–101

Chaitow L (2013) Muscle energy techniques, 4th edn. Churchill Livingstone, Edinburgh, pp 94, 243, 245, 249, 250

Chapleau J, Canet F, Petit Y, Laflamme GY, Rouleau DM (2011) Validity of goniometric elbow measurements: comparative study with a radiography method. Clin Orthop Relat Res 469(11):3134–3140

Faqin AI, Bedkar N, Shyam A, Sancheti P (2019) Effects of MET on pain, ROM and function in patients with post-surgical elbow stiffness; aA randomized control trial. Hong Kong Physiother J 25–33

Fryer G (2011) Muscle energy technique: an evidence informed approach. Int J Osteopath Med 14(1):3–9

Fryer G, Ruszkowski W (2004) The influence of contraction duration in MET applied to the atlantoaxial joint. Int J Osteopath Med 7(2):79–84

Kisner C, Colby LA (2012) Therapeutic exercise foundation and techniques, 6th edn. Jaypee Brothers, New Delhi, India

Mac Dermid JC, Vincent JI, Kieffer L et al (2012) A survey of practice patterns for rehabilitation post elbow fracture. Open Orthop J 6:429–439

Maheswari, Mahaskar (2015) Essential orthopedics, 5th edn

Matthew Hoffman MD (2021) Human anatomy. Web med @2014.LLC

Norkin CC, White DJ (n.d.) Measurement of joint motion, a guide to goniometry, 4th edn

Phadke A, Bedekar N, Shyam A, Sancheti P (2016) Effect of muscle energy technique and static stretching on pain and functional disability in patients with mechanical neck pain: a randomized controlled trial. Hong Kong J Physiother 35:5–11

Prince DD, McGrath PA, Raffi A, Buckingham B (1983) The validation of visual analog scales as ratio scales measures for chronic and experimental Pain. Pain 17:45–56

Schleip R (2003) Fascial plasticity—a new neurobiological explanation. Part 2. J Bodyw Mov Ther 7:104–116

Ylinen J (2008) Stretching therapy; for sports and manual therapies. Section-I-stretching theory, 1st edn. Oxford, United Kingdom, Churchill Livingstone, pp 22–102

Chapter 6
Computer Interactions for Sports Performance Analysis: A Review

Manjeet Singh, Malook Singh, Harish Kumar, and Paramvir Singh

1 Introduction

Several studies were carried out in 2D/3D sequence, and these examinations gathered large-scale datasets, especially on 2D images to find out HPE (Trumble et al. 2017). Substantial learning models are a type of machines that can become effective at grading by developing high-level features from low-level structures, which will mechanizing the character processing (Hinton and Salakhutdinov 2006; Hinton et al. 2006; Bengio 2009). The issue of human posture recognition can be effortlessly settled by utilizing a deep convolutional neural network. A convolutional neural network (CNN), a type of deep learning system, investigates an input image and provides distinct aspects of the image and different weights in order to distinguish them. There are a few steps involved, for instance; the initial processing phase uses low-cost CNN in comparison with other classification algorithms. Convolutional networks are trained in subsequent phases, so that they can learn the properties of incoming data, which results in a diagrammatic representation of the CNN that is comparable to the connectivity pattern of brain neurons (Krizhevsky et al. 2012). It was a feed-forward neutral network with deep assembly and convolutional calculation built by imitating the mechanism of human visual perception. In order to ensure translation invariance, CNN has the ability to extract attributes and categorize the input instruction based on its graded structure. The review was giving the information on various human body models and datasets that are utilized in human posture assessment procedures. Nowadays, deep learning was a wide area for learning human postures and sports. During earlier studies, i.e., before 2016, the focus is on the picture models and exemplar-based models. But during recent times, quick progress of deep

M. Singh · M. Singh · H. Kumar · P. Singh (✉)
Department of Sports Science, Punjabi University Patiala, Patiala 147002, India
e-mail: tparamvir@gmail.com

M. Singh
e-mail: manjeetgaggon@gmail.com

A. D. Singh et al. (eds.), *Sports Ergonomics*, Design Science and Innovation,
https://doi.org/10.1007/978-981-97-7804-1_7

learning methods (3D pose estimation) are introduced. It has been studied that 3D pose estimation had gained mindfulness in the area of a computer algorithm.

2 Human Body Model, Datasets, and Evaluation

Due to the complex structure of the human body, more than one type of model is considered. The basic and simple model used for pose estimation is the skeleton-based model. All human body models are as follow.

SMPL-Based Model: Skinned multi-person linear model was one of the models of human body model (Loper et al. 2015), as shown in Fig. 1. SMPL is a skinned vertex-based model in which wide variations of the body structure of human poses are perfectly measured. The human skin was expressed as a triangulated web, in which the shape and pose parameters are observed. The 3D pose location will be estimated by studying the body and shape parameters.

Skeleton-Based Model: The first model that was commonly used for 2D human posture estimation (Cao et al. 2018) is shown in Fig. 2. The human skeleton model was represented as a tree structure that contains key points of that body and connects adjacent joints using edges between key joints.

Surface-Based Model: A recently used model in human posture estimation and is named "dense pose" (Alp Güler et al. 2018). While examining the reality that minimum scattering of key points was not enough for image recognition or present statue of the body. To counter the matter, a new dataset called dense pose-COCO is established. This work advances the evolution of human posture understanding in different images and can be recognized as another crucial move in the field of works

Fig. 1 SMPL model (Loper et al. 2015), the white points are predefined

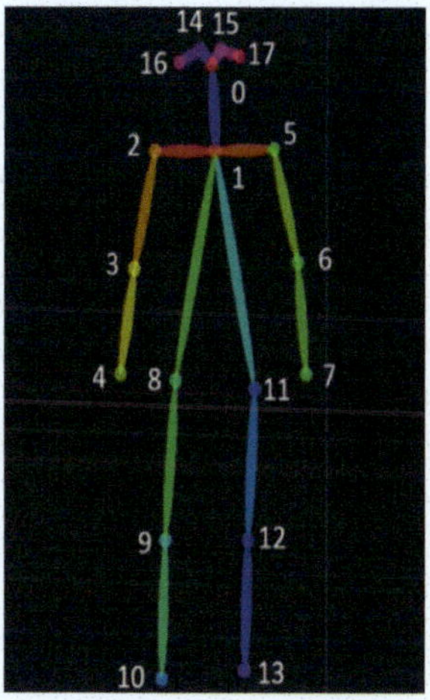

Fig. 2 Skeleton-based model key points

on exceeding the different standards for humans in 2D or 3D human estimation datasets, such as MP II Human Pose (Andriluka et al. 2014), Microsoft COCO (Lin et al. 2014), HumanEva (Sigal et al. 2009), Human3.6 M (Ionescu et al. 2013).

3 Datasets

The human pose estimation datasets were mainly collected based on a motion capture system. There are different types of datasets used to analyze various images or videos during recent years. The HumanEva dataset (Sigal et al. 2009) and Human 3.6 M dataset (Ionescu et al. 2013) were the most used datasets for posture estimation. Some of the 3D datasets used for the recognition of human postures are-

Human 3.6 M Dataset (Ionescu et al. 2013) was the largest dataset with the correct 3.6 million 3D poses. These 3D poses are obtained from six men and five women as they perform various natural poses and movements such as posing, hailing, eating, and talking on the phone and then put them under an organized image, human motion capture, and 3D body scan of different participants. This dataset involves different activities which provide accurate poses and positions of the different joints of human postures.

HumanEva-I Dataset (Sigal et al. 2009) was carried data from four subjects that perform a given set of actions that are predefined. These sets are executed in three repetitions (two times with video and motion capture, and one with motion capture alone). The HumanEva-I consist of seven videos of different sequence such as four grayscale and three colors. The datasets will support the growth of new human motion and posture estimation algorithms. This dataset provides a criterion for the assessment and differentiation of upcoming methods.

CMU Panoptic (Joo et al. 2017) used about eight subjects for a different posture with approx. 1.5 million images. They used 480 synchronized VGA cameras, 31 HD cameras were organized in their studio, and 10 RGB-D sensors were used for capturing human motion. All of the 521 cameras were calibrated by structure from the motion approach.

SURREAL (Synthetic hUmans foR REAL tasks, (Varol et al. (2017)) was the largest-scale synthetic dataset with some randomly generated 3D poses, shapes, textures, illustrations, and backgrounds. This dataset provides the information of body segments and depth estimation. For 3D motion capture markers, the motion and shape capture method were used for SMPL parameters (Loper et al. 2014). Then, by proper parameters and SMPL model, the synthetic body was created, and the actual and real surfacing image was mapped into the body shape. The highly increased validity of the synthetic data was carried by the quality information that was carried out from 3D scans of the various person wearing normal clothing. The dataset which provides the first 3D pose annotation, part segment, and ground depth estimation was SURREAL. Lastly, out of all the subjects, 115 participants were used for the training, and 30 of them performed the test set.

MPI-INF-3DHP (Mehta et al. 2017) is a dataset that includes multiple camera systems to collect the data for posture estimation. During the data collection, the dataset consists of eight subjects (4 male + 4 female) who perform various activities such as sitting, walking, complex exercise poses, and dynamic actions. The MPI-INF-3DHP covers more poses than Human 3.6 M. No particular type of clothes was needed for performing different poses. The activity of posing almost covers the time period not more than 1 min. During the process, the subject put on two sets of clothing. One set of clothing was casual style and the other was plain colored. In order to raise the variety of data, each subject performs activities of both daily apparel and plain-colored clothing sets. Additionally, the dataset increased the value of ground truth position and contextual augmentation by providing Chroma-key masks for the background.

JTA (Joint Track Auto, Fabbri et al. (2018)) was a manufactured individual tracking dataset in a metropolitan framework used for ground truth interpretation of 3D poses, of which 512 videos are used for training (256) and testing (256). The collected videos with various luminescence conditions and frames of reference are taken from the highly photorealistic video games GTA V by Rockstar North. The variation distance was 0.1 to 100 m from cameras and the heights of related participants are different from 20 to 1100 pixels. The examination of the game rendering module includes 14 body parts that are automatically interpreted in Andriluka et al. (2014, 2018). In addition to that, there are numerous simulated obstacles including occlusion and self-occlusion which are flagged. If every joint is not directly visible to the camera view point, then it is pronounced occluded, and if the joint belong to the same person then it is self-occluded. Besides, for re-identification research, every person was allocated an identifier, so that the dataset was formed in an appropriate manner.

KTH Multiview Football Dataset II (Burenius et al. 2013) was a dataset which consists of different images of football players during a match of the Allsvenskan league. While collecting the data, two different parts were introduced: first in 2D along with ground truth pose and second in both 2D and 3D with ground truth pose. In 3D dataset, two football players were involved, and two different sequences for each player and about 2400 images from three different views along with 800 time frame were captured. For each frame of 14 annotated joints, different views are modified and organized in relation to 3D ground truth pose along with orthographic camera matrix. In 2D dataset, three football players with 5907 different images along with 2D ground truth pose for each image and 14 annotated joints were involved.

3DPW (3D Poses in the Wild, von Marcard et al. (2018)) was the dataset which includes more than 51,000 frames with accuracy 3D pose in the wild, while walking around the city, having coffee or getting the bus. The data of images was collected by using hand-held phone camera. In order to get the actual IMU readings and to gather the images of all frames in video arrangement 3D poser named video inertial poser (VIP) was used. VIP was valid video poser with 26 mm of accuracy, and this accuracy was enough for the formation of dataset for 3D pose estimation. Different number of IMUs are used for tracking different subjects such as for single subject tracking, 17 IMUs were used, while 9–10 IMUs were used to track two different subjects. Then, the video and IMUs data are organized by a clapping motion as in Pons-Moll et al. (2011). In total, the dataset consists up to 18 different clothing styles and movement actions such as walking in cities, going up-stairs, having coffee, or taking the bus. Compared with total capture, there are more subjects in a scene (Table 1).

4 Discussion

3D analysis of the human body generally takes the task of predicting the postural articulations which enable to produce human movement patterns. It was a research activity that includes a wide range of human posture and its fundamental movement activities with the involvement of a computerized algorithm. In the present review, it was an attempt to locate and drive different research works in the area of deep learning techniques with the help of research paper published in the journals of repute that enables the research activity to progress and also try to initiate an idea to study 2D, 3D, static and dynamic images for future analysis. No doubt, the 3D human posture estimation provides additional information as compared to 2D posture estimation. But the inclusion of static as well as dynamic image reference gives an additional wide scope of applications. The application part for all illustrations possibly makes a distinction to significantly modulate image as well as video identification task make more accuracy.

Generally speaking, the application part with 3D posture estimation will help the people working in the areas of human computer interaction, autonomous driving models, video surveillance, biomechanics, sports education and performance, fashion

Table 1 Different datasets and their characteristics

Year	Dataset	No. of images	No. of subjects	Characteristics
2010	HumanEva-I	12 sequence	4 subjects, 6 actions	Indoor multiview video; marker less motion capture
2013	Human3.6 M	3.6 M	11 (5 female + 6 male)	One of the largest motion capture dataset; multiple views
2013	KTH multi-view football dataset II	In 3D (2400) In 2D (5907)	In 3D subjects (3) In 2D subjects (2)	Outdoor multiview video and cameras capture dataset for various joint and body movement in football field
2016	CMU Panoptic	1.5 M	Up to 8 subjects	Captured in a studio with hundreds of cameras; large social interaction
2016	MPI-INF-3DHP	1.3 M frames	8 (4 female + 4 male)	Indoor multiview video, markerless motion capture, data augmentation
2017	SURREAL	6 M frames	145	Rendered from 3D sequences of motion capture data (Human3.6 M)
2018	JTA	500 K frames	> 21	Massive simulated dataset, 500 K frames with almost 10 million pose
2018	3DPW	60 sequences	7	The only promising 3D pose in the wild dataset; 24 train, 24 test, 12 validation

Source: Wang et al. (2021)

designing, and psychological utilization (Du et al. 2019; Kim et al. 2019). Different studies have different significant areas, which explore human posture and motion. Posture estimation and human bodily movements tell about the health condition of different bodies by monitoring the walking, sitting, and running activities. In sports, the HPE was used to avoid injuries by explaining the correct posture technique to athletes (Airò Farulla et al. 2016). Moreover, self-operated clips of 3D poses from videos are very useful for the athlete to correct their mistakes for further training programs (Hwang et al. 2017). Fashion designing due to online shopping becomes more popular. By the use of 3D based virtual structure, anyone can try clothes according to their choices (Pons-Moll et al. 2017; Han et al. 2018).

5 Conclusions

Considering the presently available literature in terms of HPE and SPE, it had been adhered that computer interactions interface through a proper algorithm was an advanced program which will help to analyze the human body testing. Several approaches had been made using 2D/3D analyses, and also a number of models are trying to develop different systems to access the human body abilities in general and also sports specific. Different tasks representing different comparative models using

DNN, CNN, RNN, ANN, etc., in progress to recognize the focused work on action recognition, implementing the basic network, markers for additional information, and better results to enable datasets and AI introducing for the final center of the body as inputs and outputs as well. It had been concluded that this type of research approach will enable computer applications in appropriation to provide human movements and motion analyses for specific contexts as well as also for specific sports. Definitely, in the future, this method will implement to solve analytical problems in challenging environmental situations.

References

Andriluka M, Pishchulin L, Gehler P, Schiele B (2014) 2D human pose estimation: new benchmark and state of the art analysis. In: Proceedings of the IEEE conference on computer vision and pattern recognition (CVPR), pp 3686–3693

Andriluka M et al (2018) PoseTrack: a benchmark for human pose estimation and tracking. In: Proceedings of the IEEE conference on computer vision and pattern recognition (CVPR), pp 5167–5176

Badiola-Bengoa A, Mendez-Zorrilla A (2021) A systematic review of the application of camera-based human pose estimation in the field of sport and physical exercise. Sensors 21, p 5996

Bengio Y (2009) Learning deep architectures for AI. Found Trends Mach Learn 2(1):1–127

Cao Z, Simon T, Wei SE, Sheikh Y (2018) Realtime multi-person 2D pose estimation using part affinity fields. Proceeding 30th IEEE confernce computer vision pattern recognition, CVPR 2017, pp 1302–1310

Fabbri M, Lanzi F, Calderara S, Palazzi A, Vezzani R, Cucchiara R (2018) Learning to detect and track visible and occluded body joints in a virtual world. In: Proceedings of the European Conference on Computer Vision, ECCV. pp 430–446

Airò Farulla G, Pianu D, Cempini M, Cortese M, Russo LO, Indaco M, Nerino R, Chimienti A, Oddo CM, Vitiello N (2016) Vision-based pose estimation for robot-mediated hand telerehabilitation. Sensors 16(2), 208

Burenius M, Sullivan J, Carlsson S (2013) 3D pictorial structures for multiple view articulated pose estimation. In Proceedings of the IEEE conference on computer vision and pattern recognition, pp 3618–3625

Du X, Vasudevan R, Johnson-Roberson M (2019) Bio-lstm: a biomechanically inspired recurrent neural network for 3-d pedestrian pose and gait prediction. IEEE Robot Autom Lett 4(2):1501–1508

Fukushima K (1980) Neo cognition: a self-organizing neural network model for a mechanism of pattern recognition unaffected by shift in position. Biol Cyb 36:193–202

Gonzalez TF (2007) Handbook of approximation algorithms and metaheuristics, pp 1–1432

Güler RA, Neverova N, Kokkinos I (2018) DensePose: dense human pose estimation in the wild. In: Proceedings of IEEE Computer Social Conference Computing Vision Pattern Recognition, pp 7297–7306. https://doi.org/10.1109/CVPR.2018.00762

Han X, Wu Z, Wu Z, Yu R, Davis LS (2018) Viton: an image-based virtual try-on network. In: Proceedings of the IEEE Conference on Computer Vision and Pattern Recognition, pp 7543–7552

Hinton GE, Salakhutdinov RR (2006) Reducing the dimensionality of data with neural networks. Science 313(5786):504–507

Hinton GE, Osindero S, The Y (2006) A fast learning algorithm for deep belief nets. Neural Comput 18:1527–1554

Hwang J, Park S, Kwak N (2017) Athlete pose estimation by a global-local network. In: Proceedings of the IEEE Conference on Computer Vision and Pattern Recognition Workshops, pp 58–65
Ionescu C, Papava D, Olaru V, Sminchisescu C (2013) Human3.6M: large scale datasets and predictive methods for 3D human sensing in natural environments. IEEE transactions on pattern analysis machine intelligence
Joo H, Simon T, Li X, Liu H, Tan L, Gui L, Banerjee S, Godisart T, Nabbe B, Matthews I, et al (2017) Panoptic studio: a massively multiview system for social interaction capture. IEEE Trans Pattern Anal Mach Intell 41(1):190–204
Kim W, Ramanagopal MS, Barto C, Yu M-Y, Rosaen K, Goumas N, Vasudevan R, Johnson-Roberson M (2019) Pedx: benchmark dataset for metric 3-D pose estimation of pedestrians in complex urban intersections. IEEE Robot Autom Lett 4(2):1940–1947
Krizhevsky A, Sutskever I, Hinton GE (2012) Imagenet classification with deep convolutional neural networks. In: Advances in Neural Information Processing Systems, pp 1097–1105
LeCun Y, Bottou L, Bengio Y, Haffner P (1998) Gradient-based learning applied to document recognition. In: Proceedings of the IEEE 86(11):2278–2324
Lin T-Y et al. (2014) Microsoft COCO: common objects in context. Lect Notes Comput Sci (including Subser Lect Notes Artif Intell Lect Notes Bioinformatics) 8693:740–755
Loper M, Mahmoody N, Blackz MJ (2014) MoSh: motion and shape capture from sparse markers. ACM Trans Graph 33
Loper M, Mahmood N, Romero J, Pons-Moll GJB (2015) SMPL. ACM Trans Graph 34
von Marcard T, Henschel R, Black MJ, Rosenhahn B, Pons-Mol, G (2018) Recovering accurate 3D human pose in the wild using IMUs and a moving camera. Lect Notes Comput Sci (including Subser Lect Notes Artif Intell Lect Notes Bioinformatics) 11214 LNCS, 614–631
Mehta D, Rhodin H, Casas D, Fua P, Sotnychenko O, Xu W, Theobalt C (2017) Monocular 3d human pose estimation in the wild using improved cnn supervision. In 2017 international conference on 3D vision (3DV) (pp 506–516). IEEE
Pons-Moll G, Baak A, Gall J, Leal-Taixe L, Mueller M, Seidel H-P, Rosenhahn B (2011) Outdoor human motion capture using inverse kinematics and von misesfisher sampling. In: 2011 International Conference on Computer Vision. IEEE, pp 1243–1250
Pons-Moll G, Pujades S, Hu S, Black MJ (2017) Clothcap: seamless 4D clothing capture and retargeting. ACM Trans Graph 36(4):73
Sigal L, Balan AO, Black MJ (2009) HumanEva: synchronized video and motion capture dataset and baseline algorithm for evaluation of articulated human motion. Int J Comput Vis 87(1):4–27
Trumble M, Gilbert A, Malleson C, Hilton A, Collomosse J (2017) Total capture: 3D human pose estimation fusing video and inertial sensors. In: BMVC, vol 2, p 3
Varol G et al (2017) Learning from synthetic humans. Proceeding 30th IEEE Conference Computing Vision Pattern Recognition, CVPR 2017, pp 4627–4635
Wang J et al (2021) Deep 3D human pose estimation: a review. Comput vis Image Underst 210:103225

Chapter 7
Ketogenic Diet: Physical Performance and Metabolic Health of Athletes

Tarvinder Jeet Kaur and Ruhi Grewal

1 Introduction

Nutrition is predominantly, one of the essential parameters in the adequate maintenance of athlete's performance. Absence of balanced and well to do nutrition can impact the physical and metabolic performance of sports person. While planning and governing the diet of athletes, it is extremely important to highlight and emphasize more on the restricted amount of glycogen stores in liver and muscle tissues as delaying the depletion of glycogen stores can help in preventing it is exhaustion at any point of strenuous physical activity executed by athletes (Mata et al. 2019). To avoid the issue of depletion, carbohydrates are being incorporated in the diet so as to provide fuel from the glucose which is the greatest source of energy. But it is highlighted that nowadays, athletes are currently using in demand and popular ketogenic diet (Volek et al. 2016a), where carbohydrates are immensely reduced, cut down and substituted with the fats causing the body to place in the condition of ketosis which is represented by the raised level of ketone bodies in the blood due to breakdown of non-carbohydrate sources like fats and proteins (Cox et al. 2016). Randomized controlled trial has depicted that this kind of diet utilizes fat instead of carbohydrates where beta oxidation of fatty acids provide body fuel to cells and tissues which succours athletes in gaining excellence in physical performance (Cox et al. 2016). But this well liked and in demand ketogenic diet among athletes is considered to be fad as its claim for sustainable health perpetuation with eminence in physical performance is unaccompanied by any certified dietary directions. This fad diet is deprived of very essential food group components like cereal and grains consisting of dietary fibre, carbohydrates, vitamins, minerals and antioxidants like phytochemicals which is required for execution of voluntary and involuntary body activities (Fad Diets 2021).

T. J. Kaur (✉) · R. Grewal
Department of Home Science, Kurukshetra University, Kurukshetra, India
e-mail: tarvinkhushi1974@gmail.com

A. D. Singh et al. (eds.), *Sports Ergonomics*, Design Science and Innovation,
https://doi.org/10.1007/978-981-97-7804-1_8

As per depicted by nutritional articles and blogs, ketogenic diet is of different types like traditional, standard, cyclic, targeted and keto 2.0 Traditional ketogenic diet involves the intake of 1 gram of protein and carbohydrates for every 4 grams of fats. This type of diet is suitable for therapeutic purposes like epilepsy, cancer, etc. Standard ketogenic diet entails the intake of 55–75% of fats, 15–35% of protein and 0–10% of carbohydrates where people expecting for the weight loss management pay attention to such type of diet. High protein keto diet involves 30–35% of proteins, 50–60% of fats and about 10% of carbohydrates. High proportion of protein in diet aids in intensifying and reinforcing the muscle mass which is required for all type of body builders or people executing for strength training activities. Cyclic keto diet implies for standard keto diet accompanied by refuelling of carbohydrates which can go up to 100–150 grams of carbohydrates but at specific and designated interval of times. It is also known as carb cycling where athletes executing physical activities like fast running, tennis, cycling for hours necessitate such type of diet as they require glucose for body fuel. Targeted keto diet is in between standard and cyclic where carbohydrate refuelling is done before the initiation of any physical performance so as to acquire the energy shots. Keto 2.0 emphasizes more on quality than quantity where plant based diet is focused comprising of healthy fatty acids like omega 3 and omega 6. Here, carbohydrate restriction is little eased and its proportion is fabricated to moderate level (Six types of keto diet and how to get started 2020; Ketogenic diet 2020).

Accentuating on the physical activities performed by athletes, research studies and blogs on health and sports have weighed on the variation of intensities of physical activities which has a huge dietary impact. There are basically two types of intensities of physical activities performed by athletes—(i) low intensity and (ii) high intensity physical activities. Low intense physical activities are an aerobic cardio which can be performed for a longer duration with a heart rate maintained at the steady position of 60–70%, thus making it favourable for an athlete person to perform for extended period of time easily without being exerted. Short jog, marathon running, swimming, yoga, pilates, etc., performed at a steady rate are an examples of low intensity physical activities. Whereas high intensity activity is an anaerobic type of physical activity which is a short burst of high intense physical efforts executed by athletes for short span of time on the grounds of cyclic intervals. Here, heart rate is pushed up to 70–90% accompanied by raised breathing rate. Fast running, soccer, jumping rope, cycling 10 miles per hour, tennis (singles), carrying heavy loads, etc., are all the example of high intensity exercises. Thus, it is extremely crucial for athletes to accentuate on the different intensities of physical activities while relying on the ketogenic diet (High intensity vs low intensity workouts 2021; High intensity vs low intensity Intensity Cardio 2022).

Though presently ketogenic diet is contemplated to be advantageous for athlete but researchers have ascertained that persistence of its fad quality have severe repercussions on physical performance and metabolic health of athletes by negatively impacting the intensity of activities and causing metabolic disturbances like dyslipidaemia, ketoacidosis, renal stones, dehydration, renal disease, hyperuricemia, keto flu, etc., respectively (7 side effects of going keto 2022).

So while planning and organizing the ketogenic diet for athletes, kind of physical activity performed by athletes accompanied by the proportion and quality of macronutrients to be modified and incorporated is decisive and crucial in determining the well-nourished keto diet for athletes. Keeping this in view, a present review study is conducted to deliver and portray the information about the role of ketogenic diet in physical performance and metabolic health of athletes.

2 Methodology

A review of research was performed where information related to the prominence of nutrition in dietary regimen of athlete was gathered. Variant sports articles and review papers were searched for collecting the details on importance of delaying the exhaustion of limited glycogen reserves. Facts and particulars related to the ketogenic diet was acquired from variant review and research papers on sports and nutrition using the data bases like Scopus, Google Scholar, PubMed, Crossref, Science Direct, etc. Different nutritional blogs on fad ketogenic diet were explored. Different types of ketogenic diet were listed down by acquiring the information from various nutritional blogs, websites and review papers. Physical performance of athletes were spotlighted where details related to variant intensity of physical activities like low and high intensity physical activities were collected using sports blogs and articles. Besides the physical performance, metabolic health of athletes was also emphasized by exploring various research papers published in recognised journals. Latest cross sectional, meta-analysis, clinical trials and systematic reviews were utilized for the literature collection about the biochemical process of ketogenic diet. Clinical trials, blogs and articles on sports and nutrition were purposely selected so as to obtain information about role of ketogenic diet on different intensities of physical activity and metabolic health of athletes causing several health disturbances. Search terms like 'glycogen stores', 'ketogenic diet', 'low intensity activity', 'high intensity activity', 'aerobic exercises', 'anaerobic exercises', metabolic health, 'dyslipidaemia', 'impact of ketogenic diet on athletes', 'ketogenic diet and athletes', 'ketogenic diet and high endurance physical activities', 'ketogenic diet and dyslipidaemia', 'ketogenic diet and renal diseases', 'ketogenic diet' and keto acidosis', 'ketogenic diet and metabolic disturbances', etc., were used for the review study.

3 Results

Intake of a normal balanced diet comprised of balanced macronutrients goes through a biochemical process where beta oxidation of stored carbohydrates and fats releases glucose and fatty acids, respectively, which sequentially is converted to 2 molecules of pyruvate and then it is transformed to acetyl CoA in the presence of enzyme HS-CoA. Acetyl CoA formed combines with oxaloacetate which in turn enters the

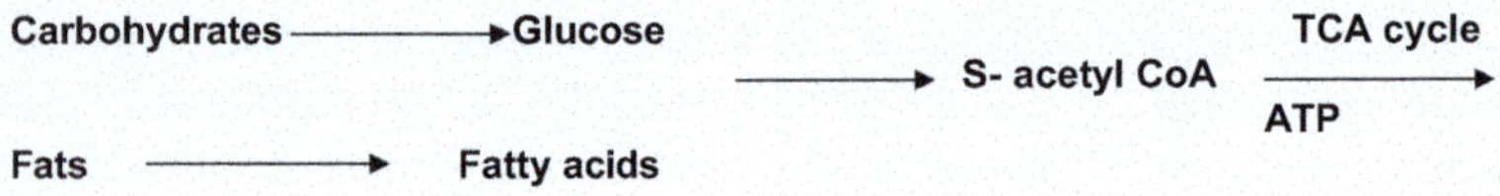

Fig. 1 Metabolism of carbohydrates

Kreb cycle and produces the ATP, thus releasing the carbon dioxide and water as illustrated in Figure 1 (Carbohydrate metabolism 2017).

But in the case of low carbohydrate and high fat ketogenic diet, review studies highlighted that acetyl CoA formed from glucose or fatty acids doesn't enter the Kreb cycle for ATP production. Besides that, it is converted to acetoacetate or 3-hydroxybutyric acid chemical components which are known as ketone bodies as shown in Figures 2 and 3.

These ketone bodies function as main source of body fuel dominantly derived from the oxidation of fats by mobilizing and advancing the raised level of fatty acids (Volek et al. 2016a; Goodridge and Sul 2000). Review researchers have ascertained that intake of low carbohydrate and high fat ketogenic diet alters the casual cyclic pathway of glucose metabolism to high fat dependant biochemical process where ketone bodies formed from the circulatory fatty acids remain preserved and cannot get into the account of body cells and tissues for the acquirement due to deprivation of mitochondrial enzyme. In that case, ketone bodies progresses from hepatic tissue to extra hepatic tissues and actuates the blood–brain fencing shipper which is chargeable for raising the concentration of ketone bodies by increasing the rate of blood flow. This elevated concentration of ketones has the functional ability in working as an alternative to glycogen in meeting the unrestricted and unhindered energy needs to

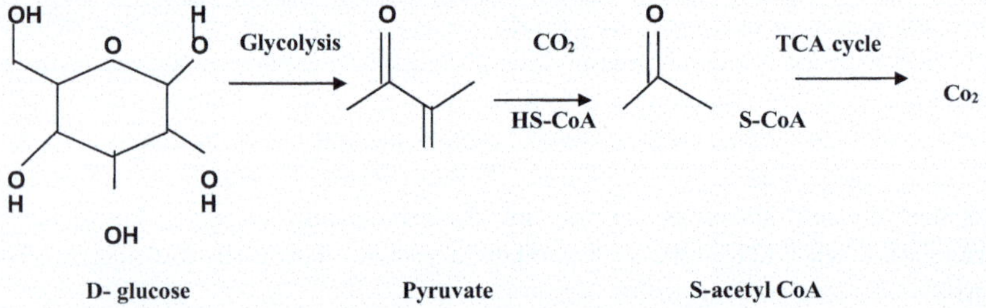

Fig. 2 Metabolism of carbohydrates

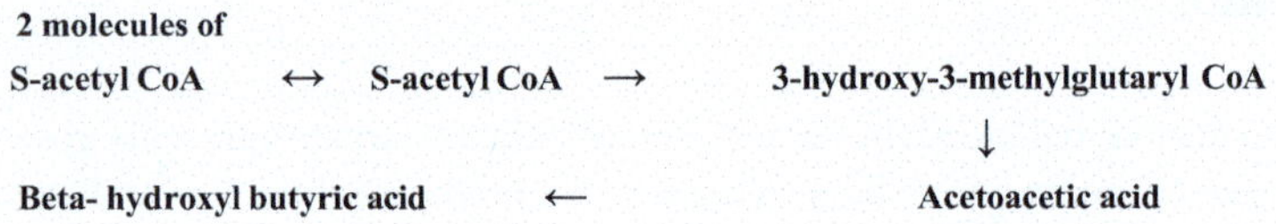

Fig. 3 Metabolism of carbohydrates

different cells and tissues at the high rate of speed. Raised fat, protein and low carbohydrate ketogenic diet restricts the cells and tissues for the uptake of glucose and reprieves the process of glucose metabolism by substituting glucose molecules with ketone bodies which is chargeable for lessening the activity of glycolysis (catabolization of glucose to pyruvic acid and energy in the presence several enzymes) (Watford and Goodridge 2000; Salway 1999).

4 Ketosis and Ketoacidosis

Research trials have stated that intake of ketogenic diet produces ketone bodies and create the condition of ketosis where the concentration of ketone bodies in the blood is elevated. On the comparing the normal diet with ketogenic diet, it was observed that normal diet produces ketone bodies with a range less than of 0.3 mM, whereas ketogenic diet creates the condition of ketosis with a range between 0.5 and 3 mM. It is extremely crucial for ketogenic diet to sustain and balance that range of ketosis which succours in providing efficient body fuel for performing physical activities. The condition of ketosis created by the ketogenic diet is extremely different from diabetic ketoacidosis where level of ketone bodies is elevated to greater than 10 mM which sequentially is accountable for producing the acidic environment in body by destabilizing the acid-base buffer systems. The excessive elevated range of ketone bodies forbade and prevent the further production of ketogenesis which in turn is responsible for averting the process of lipolysis, beta oxidation and ketogenesis (Salway 1999; Zammit 2001; Björntorp 1996).

5 Physical Performance and Ketogenic Diet

Ketogenic diet has an utmost part in the physical performance of athletes. 100 g of combustion of 3-hydroxybutyrate and acetoacetate produces about 10.5 kg and 9.4 kg of ATP, respectively, which has an extreme role to play in the type of intensity of physical performances executed by athletes. It is immensely crucial to select the kind of physical intensity favourable and suitable for athletes who are relying on high fat and low carbohydrates ketogenic diet (Salway 1999).

(1) Low intensity physical activities and ketogenic diet:

Low intensity or aerobic physical activity succours in modifying the body to ketogenic diet where the intake of carbohydrate is extremely reduced and proportion of fat is elevated. Diverse clinical trials conducted on physical performance of athletes revealed that by advancing fat as a fuel rather than carbohydrate, ketogenic diet is considered to be favourable for aerobic exercises. Low intense physical activity like endurance training executed by athletes aids in adapting the body to the functional capability of fat oxidation where metabolic pathways of beta oxidation of fatty acid strengthens and intensifies with the elevation in duration of physical performance from low to maximum (McSwiney et al. 2018; Egan and Zierath 2013). Research studies highlighted that trained athletes performing physical activities have higher mitochondrial or capillary density which in turn is accountable for successfully raising the oxidation of fats towards muscle cells and tissues. Apart from this, athletes dependant on aerobic exercise have the operational capability in sparing the glucose metabolism successfully without being exerted or fatigue where glycogen stores are not impacted (Goto et al. 2007; Goedecke and Lambert 2003). Low intense activity demands more blood and oxygen to different cells and tissues which in turn elevates the need of cardiac output and lactate production. These types of physiological changes adapt the body to the beta oxidation of fatty acids for body fuel which in turn makes athletes favourable and acceptable for ketogenic diet (Helge 2002; Cox et al. 1996). Besides the physiological adaptations, relying on low carbohydrate diet increases secretion of hormone sensitive lipase and depresses production of insulin which aids the body to modify for the condition of dietary ketosis during the execution of physical performance (Yeo et al. 2011). Research and review studies ascertained that athletes relying on ketogenic diet achieve appreciable results in the aerobic activities like short jog, marathon running, swimming, yoga, pilates, etc., when performed at a steady rate (Table 1).

(2) High intensity physical activity and ketogenic diet:

Table 1 Following research studies showcasing about the advantageous impact of ketogenic diet on low intense physical activities

Type of studies	Type of diets	Outcomes	References
Clinical Trial: 20 male endurance-trained athletes were fed with diet for 12 weeks	Low carbohydrate ketogenic diet (LCKD) and carbohydrate based diet	LCKD improved body composition and utilization of fat oxidation during the physical performance	Mcswiney et al. 2018)
Interventional study: Two groups of 20 elite ultra-marathoners and ironman distance tri athletes participated in study for 20 months	High and low carbohydrate diet, respectively	Low carb diet Ironman distance tri athletes have high rate of fat oxidation	Volek et al. 2016b)
Clinical Trial: 11 crossfit trained males and female cyclists each consumed diet for 4 weeks	Ketogenic based diet	Improvement in fat utilization during the submaximal intensity physical performance	Durkalec-Michalski et al. 2019)
Randomized controlled trial: 8 trained male endurance runners were fed with diet for 31 days	Normal diet and isoenergetic ketogenic diet	Keto diet-Enhanced submaximal physical performance functional ability	Shaw et al. 2019)

Numeral interventional studies performed by researchers showed up that efforts loaded by athletes performing high intense physical activities require fuel which dominantly depends upon skeletal muscle and glycogen system where energy needs are met by lactic acid and phosphate system. Operational capability of skeletal muscles requires amino acids for strengthening and building up of muscles which can only be fulfilled by high protein diet. Ketogenic diet is favourable as it succours in meeting the nutritional needs of athletes by providing more than 15% of proteins. But supply of protein from ketogenic diet has a huge metabolic flaw as the persistence of low carbohydrate in diet inflames the protein metabolism. Relying on ketogenic diet raises the process of gluconeogenesis which is accountable for providing glucose to cells and tissues derived from amino acids. So amino acids mainly are consumed in providing body fuel, whereas the dominant function of protein in strengthening the muscles tissue gets worsened. Besides this, numeral clinical trials have depicted that elevated consumption of amino acids for body fuel has shown the lessened power strength and extreme weight loss depicting zero absolute power among athletes performing anaerobic exercises and relying on ketogenic diet (Gregory et al. 2017; Wilson et al. 2020). Decreased muscle power, enfeeble strength gain and elevated muscle hypertrophy were also observed among athletes performing resistance training. Repercussions have been observed on ATP production where research studies have highlighted that ATP released from the oxidation of fats during the consumption of ketogenic diet cannot be toned up with ATP released from anaerobic exercises as ATP production during high intense exercises is at an elevated range which becomes strenuous for athletes who usually depend upon dietary ketone for body fuel (Wroble et al. 2019; Body et al. 1974). To achieve high calories for the execution of anaerobic physical activity, body entails ketone bodies which usually depend on high or frequent intake of ketogenic diet which causes excessive accumulation of blood ketones in the blood leading to fatigue or lassitude during the physical performance. Clinical trials and review studies have represented that metabolic flaws observed during high intensity physical performance accompanied by intake of ketogenic diet have severe consequences on physical performance of athletes (Table 2).

Besides affecting the intensities of physical performances, ketogenic diet has critical consequences on the metabolic health of athletes by giving rise to several metabolic disruptions like dyslipidaemia, hypoglycaemia, ketoacidosis, renal stones, dehydration, renal disease, extreme weight loss, hyperuricemia, etc.

6 Metabolic Health

As per depicted by interventional studies, athletes relying on ketogenic diet have been found to suffer from metabolic disturbances like dyslipidaemia. Nutritional content and composition of ketogenic diet needs to be highlighted as it ruins the metabolic health of athletes by causing the issue of perturbed lipid profile, where high proportion of fats, moderate or high animal protein and low carbohydrates is scientifically

Table 2 Following research studies highlighting about the ramifications of high intense physical activities among athletes while relying on ketogenic diet

Type of studies	Objectives	Outcomes	References
Clinical trial: 8 athlete gymnasts where keto and western diets were fed to them	To determine strength capacity	No significant difference in strength test was observed	Paoli et al. 2012)
Interventional study: 12 athletes were intervened with keto diet for duration of 12 weeks	Body composition was determined	DXA lean leg mass was reduced by 1.4%	Kephart et al. 2018)
Randomized controlled trial: 16 exercise trained men and women were fed with low carbohydrate ketogenic diet for duration of 4 days	To assess the anaerobic physical performance of athletes	High intensity-short duration physical performances were weakened	Wroble et al. 2019)
Symposium review: Different studies using data bases like Google Scholar, PubMed, Scopus, etc. were used	To ascertain the impact of ketogenic diet on high intensity performance	Glycogen stores get negatively impacted, showing repercussions on the economical energy	Burke 2020)

liable for elevating the lipid parameters like total cholesterol, LDL cholesterol and triglycerides levels. Athletes acquire body fuel from fat and protein which is dominantly derived from saturated fat and animal protein (Dashti et al. 2006). Thus, it becomes extremely crucial to modify the quality of macronutrients by substituting the excess of saturated fats and animal protein with unsaturated fats and plant proteins in dietary life style of athletes, respectively. Besides causing dyslipidaemia, ketogenic diet is responsible for engendering the metabolic disturbance of ketoacidosis. Healthy or diabetic athletes fully relied on ketogenic diet results into the condition of dietary ketosis which sequentially can trigger the issue of ketoacidosis where ketone bodies get excessively accumulated in the blood which in turn produces the acidic environment causing the ruination of liver, kidneys and brain and if persisted unrestrained it can become fatal (Yurkewicz et al. 2017). Therefore, it is scientifically derived that frequency or duration of ketogenic diet to be restricted as per the accessibility of macronutrients towards the metabolism of athletes. Apart from it, extreme reduction of carbohydrates in diet brings about escalated water excretion from the body as carbohydrates are reserved with water comprising of several electrolytes like sodium, potassium, magnesium and its excessive excretion elevates the elimination of electrolytes from body which in turn is chargeable for causing the complication of dehydration and electrolyte disparity. This complication engenders to the problem of *keto flu* which is depicted by several symptoms like fatigue, headache and muscle cramps which sequentially have serious repercussion on the physical performance of athletes (Bostock et al. 2020). So while planning the ketogenic diet for athletes, quantity and quality of carbohydrates to be spotlighted as its immense restriction can cause the problem of *keto flu*. Furthermore, scientific studies also ascertained that low carbohydrate and high protein diet give rise to the issue of renal disease where high protein diet lays freight on the functional capability of kidneys by excreting excess of nitrogen in the urine, whereas under normal diet, kidneys excrete proteins in a very less proportion which in turn engenders scarce amount of nitrogen in urine. Relying on ketogenic diet can cause the problem of proteinuria indicating the presence of excessive protein in the urine. Hence, proportion of proteins to be kept in a moderate ratio as proteinuria give rise to the renal disease where kidneys have to work hard to excrete nitrogen which successively causes several issues like unrestricted urination, swelling in hands, foamy or frothy urine, muscle cramps, etc (Crossby et al. 2021). Accompanied by renal disease, systematic review and meta-analysis depicts that ketogenic diet is also liable to increase the risk of renal stones where high fat diet increases the metabolism of fat which elevates the amount of acid in the blood. Acid load in the blood raises the excretion of calcium which in turn changes the biochemistry of urine indicating high level of uric acid and calcium consequently combines to form renal stones (Acharya et al. 2021). Raised acid level has also reduced the process of bone synthesis by leaching out of calcium from bones and elevating the calcium excretion level (Heikura et al. 2020). Elevated macronutrients in ketogenic diet also engendered to the problem of hypothyroidism where T3 levels are decreased due to immense calorie reduction derived from carbohydrates. Lessened T3 levels declines the metabolism which is an essential operational capability among athletes. This causes diminished athletic performance, fatigue, muscle camps, dry skin, constipation, etc (Lacovides et al. 2022). Hence, it is derived and deduced that while planning the ketogenic diet for athletes, proportion and quality of macronutrients to be focused and enhanced so as to avoid the metabolic disturbances among athletes (Table 3).

7 Conclusion

Ketogenic diet portrays an eminent role in the physical performance and metabolic health of athletes. It is important to reserve the limited amount of glycogen present in the muscle and liver tissues which can only be possible by prioritizing the type of intensities of physical activities while relying on ketogenic diet. Several interventional and review studies governed that redirecting from high intensity to low intensity physical activities aid the athletes in achieving appreciable and favourable results of physical performance as low intense activities reserves the restricted proportion of glycogen for future and utilizes the fatty oxidation for body fuel. Apart from physical performance, it is scientifically observed through different clinical trials that metabolic health of athletes can be ruined by causing the issue of dyslipidaemia, renal disease, ketoacidosis, renal stone, etc. Hence, it is eminent to emphasize on the quality and quantity of macronutrients of ketogenic diet while planning the diet for athletes. This type of physical and dietary modification can aid in achieving appreciable and progressive results by enhancing the physical performance of athletes and reducing the metabolic disruptions.

Table 3 Research and review studies have been listed down which depicts the metabolic disturbances caused by ketogenic diet on athletes

Type of studies	Objectives	Outcomes	References
Interventional study: 20 male ultra-endurance runners where high carbohydrate and low carbohydrate keto diet were consumed	To determine the impact of ketogenic diet on blood lipid parameters	Elevation in total cholesterol, LDL and HDL cholesterol	Creighton et al. 2018)
Randomized cross over metabolic study: 15 healthy subjects were fed with keto diet for 1 week	To determine the impact of high fat diet on the health status	Increase in the occurrence of kidney stones causing acidosis	Tracy et al. 2014)
Review study: Derived from recognised journals using different databases	To ascertain the role of keto diet on non-alcoholic fatty liver disease (NAFLD)	High incidence NAFLD was observed	Perdomo et al. 2019)
Interventional study: 12 healthy athletes where keto ester diet was fed	To determine the impact of ketoacidosis	Elevation in the range of d-β-hydroxybutyrate which sequentially dropped the blood pH to 7.37	Dearlove et al. 2019)

References

Acharya P, Acharya C, Thongprayoon C et al (2021) Incidence and characteristics of kidney stones in patients on ketogenic diet: a systematic review and meta-analysis. Diseases 9(2):39

Björntorp P (1996) Effects of ketone bodies on lipolysis in adipose tissue in vitro. J Lipid Res 7:621–626

Body A, Giamber S, Mager M et al (1974) Lactate inhibition of lipolysis in exercising men. Metabolism 23:531–536

Bostock ECS, Kirkby KC, Taylor BV et al (2020) Consumer reports of "Keto Flu" associated with the ketogenic diet. Front Nutr 7(113):20

Burke LM (2020) Ketogenic low carbohydrate, high fat diet: the future of elite endurance sport? J Physiol 599(3):819–843

Carbohydrate metabolism (2017). https://researchgate.net/publication/321938058_Carbohydrate_metabolism. Accessed: 10 Oct 2022

Cox C, Brown R, Mann J (1996) The effects of high carbohydrate versus high fat dietary advice on plasma lipids, lipoproteins, apolipoproteins and performance in endurance trained cyclists. Nutr Metab Cardioavscular Dis 6:227–233

Cox PJ, Kirk T, Ashmore T et al (2016) Nutritional ketosis alters the fuel preferences and thereby alters the fuel preferences and thereby endurance performance in athletes. Cell Metab 24:256–268

Creighton, BC, Hyde PN, Maresh CM et al (2018) Paradox of hypercholesterolemia in highly trained, keto adapted athletes. BMJ Open Sport Exerc Med 4(1)

Crossby L, Davis B, Joshi S et al (2021) Ketogenic diets and chronic diseases: weighing the benefits against risks. Front Nutr 16(8):702802

Dashti HM, Al-Zaid NS, Mathew TC et al (2006) Long term effects of ketogenic diet in obese subjects with high cholesterol level. Mol Cell Biochem 286:1–9

Dearlove DJ, Faull OK, Rolls E et al (2019) Nutritional Ketoacidosis during the incremental exercise in healthy athletes. Front Physiol 10(290)

Durkalec-Michalski K, Nowaczyk PM, Siedzik K (2019) Effect of a four week ketogenic diet on exercise metabolism in crossfit- trained athletes. J Int Soc Sports Nutr 16:16

Egan B, Zierath JR (2013) Exercise metabolism and the molecular regulation of skeletal muscle adaptation. Cell Metab 17:162–184

Fad Diets (2021). https://myclevelandclinic.org/health/articles/9476-fad-diets. Accessed: 10 Oct 2022

Goedecke J, Lambert E (2003) Adaptation to a high fat duet for endurance exercise: review of potential underlying mechanisms. Int J Sports Med 4(1)

Goodridge AG, Sul HS (2000) Lipid metabolism-synthesis and oxidation. Biochem Physiol Asp Hum Nutr, 305–350

Goto K, Ishii N, Mizuno A et al (2007) Enhancement of fat metabolism by repeated bouts of moderate endurance exercise. J Appl Physiol 102:2158–2164

Greene DA, Varley BJ, Hartwig TB et al (2018) A Low –carbohydrate ketogenic diet reduces body mass witout compromising performance in powerlifting and olympic weightlifting athletes. J Strength Cond Res 32:3373–3383

Gregory RM, Hamdan H, Torisky DM et al (2017) A low carbohydarate ketogenic diet combined with 6 weeks of crossfit training improves body composition and performance. Int J Sports Exerc Med Predatory 4(54)

Heikura IA, Burke LM, Hawley JA et al (2020) A short term ketogenic diet impairs markers of bone health in response to exercise. Frontiers 10:1–10

Helge J (2002) Long term fat diet adaptation effects on performance, training capacity and fat utilization. Med Sci Sports Exerc 34:1499–1504

High intensity vs low intensity Intensity Cardio (2022). https://soneva.com/high-intensity-vs-low-intensity-cardio/. Accessed: 9 Oct 2022

High intensity vs low intensity workouts (2021). https://www.breathewellbeing.in.blog/hiit-vs-low-intensity-workouts-what-you-need-to-know/. Accessed: 9 Oct 2022

Kephart WC, Pledge CD, Roberson PA et al (2018) The three month effects of a ketogenic diet on body composition, blood parameters and performance metrics in crossfit trainees: a pilot study. Sports 6(1):1

Ketogenic diet: A detailed beginner's guide to keto, Different types of ketogenic diet (2020). https://www.healthline.com/nutrition/ketogenic-diet-101#diet-types

Lacovides S, Maloney SK, Bhane S et al (2022) Could the ketogenic diet induce a shift in thyroid function and support a metabolic advantage in healthy participants? A pilot randomized-controlled- crossover trial. PloS One 17(6)

Lambert E, Speechly D, Dennis S et al (1994) Enhanced endurance in trained cyclists during moderate intensity exercise following 2 weeks adaptation to a high fat diet. Eur J Appl Physiol 69:287–293

Manninen AH (2006) Very low carbohydrate diets and preservation of muscle mass. Nutr Metab 3:1–4

Mata F, Valenzuela PL, Gimenez J et al (2019) Carbohydrate availability and physical performance: physical overview and practical recommendations. Nutrients 11(5):1084

Mcswiney FT, Wardrop B, Hyde PN et al (2018) Keto-adaptation enhances exercise performance and body composition responses to training in endurance athletes. Metabolism 81:25–34

McSwiney FT, Wardrop B, Hyde PN et al (2018) Keto- adaptation enhances exercise performance and body composition responses to training in endurance athletes. Metabolism 81:25–34

Paoli A, Grimaldi K, D' Agostino D, et al (2012) Ketogenic diet does not affect strength performance in elite artistic gymnasts. J Int Soc Sports Nutr 9 (34)

Perdomo CM, Fruhbeck G, Escalada J (2019) Impact of nutritional changes on non-alcohol fatty liver disease. Nutrients 11(3):677

Salway JG (1999) Metabolism at a glance. Blackwell Science, Oxford

Shaw DM, Merien F, Braakhuis A et al (2019) Effect of ketogenic diet on submaximal exercise capacity and efficiency in runners. Med Sci Sports Exerc 51(10):2135–2146

7 side effects of going keto (2022). https://www.heatlh.com/weight/keto-diet-side-effects. Accessed: 10 Oct 2022

Six types of keto diet and how to get started (2020). https://www.carbmanager.com/article/xtqaeh eaacmachsv/different-types-of-ketogenic-diet, accessed- 9th oct 2022.

Tracy CR, Best S, Bagrodia A et al (2014) Animal protein and the risk of kidney stones: a comparative metabolic study of animal protein sources. J Urol 192(1):137–141

Volek JS, Freidenreich DJ, Saenz C et al (2016a) Metabolic characteristics of keto-adapted ultra-endurance runners. Metabolism 65(3):100–110

Volek JS, Freidenreich DJ, Saenz C et al (2016b) Metabolic characteristics of keto-adapted ultra-endurance runners. Metabolism 65:100–110

Watford M, Goodridge AG (2000) Regulation of fuel utilization. Biochem Physiol Asp Hum Nutr, 384–407

Williams MS, Turos E (2021) The chemistry of the ketogenic diet: updates and oppurtunities in organic synthesis. Int J Mol Sci 22(10):5230

Wilson JM, Lowery RP, Roberts MD et al (2020) The effects of ketogenic dieting on body composition, strength, power and hormonal profiles in resistance training males. J Strength Cond Res 34(12):3463–3473

Wroble KA, Trott MN, Schweitzer GG et al (2019) Low carbohydrate ketogenic diet impairs anaerobic exercise performance in exercise trained women and men: randomized sequence cross over trial. J Sports Med Phys Fitness 59(4):600–607

Yeo WK, Carey BA, Spriet L et al (2011) Fat adaptation in well trained athletes: Effects on cell metabolism. Appl Physiol, Nutr Metab 36(12)

Yurkewicz M Jr, Cordas M, Zellars M et al (2017) Diabetes and sports. Am J Lifestyle Med 11(1):58–63

Zammit VA (2001) Handbook of Physiology-Section 7: the Endocrine system-Volume II: the endocrine pancreas and regulation of metabolism, regulation of ketogenesis in liver. Oxford University Press

Chapter 8
Effect of Stretching Exercise on Recovery of Quadriceps Tendinosis: A Case Study

Arpan Chattopadhyay, Subhasmita Das, Harshita Tak, Paladi Ramya, Bikram Moyra, Divik Ranjan, and B. Hemanth Naick

1 Introduction

Tendon, a connective tissue connects muscle to bone, is known to resist the tensile strength between muscle and bone (Franchi et al. 2007). Tendon pathologies including tendinopathy, acute (partial or complete tendon tear) and chronic injury/disorder are common in active athletes, the personals with intensive musculoskeletal activities, elderly persons and sedentary persons (Wu et al. 2017; Abat et al. 2017). Among these pathologies, tendinopathy accounts for half of all musculoskeletal disorders which is degenerative and non-inflammatory disorder (Andres and Murrell 2008; Fredberg and Stengaard-Pedersen 2008). If untreated, then leads to permanent tears and ruptures of the tendon (Charnoff et al. 2022). Risk factor of tendinopathy may be due to mechanical stresses, overuse/overload, insufficient/incomplete recovery time, repetitive tendon micro-trauma, sudden exposure to elevated activity, overload and inadequate healing response (Williamson et al. 2021; Steinmann et al. 2020; Aicale et al. 2018). Tendinopathy will affect the athlete with pain, shows palpation, swelling and impaired movement at the site of tendinopathy (Kaux et al. 2011). Athletes or sporting individuals who undergo extensive repetitive loading training and completion are prone to develop tendinopathy (Ackermann and Renström 2012).The aetiology of tendinopathy is beyond our comprehension, which may be due to tendon changes, damage, overload, genetic, biological and molecular mechanism factors (Canosa-Carro et al. 2022). To some extent, imaging-based diagnosis with the help of ultrasound and magnetic resonance imaging (MRI) has helped us in precise detection of tendinosis (Weinreb et al. 2014; El-Deek et al. 2019). Despite the common treatment strategies available, non-surgical options including

A. Chattopadhyay · S. Das · H. Tak · P. Ramya · B. Moyra · D. Ranjan · B. Hemanth Naick (✉)
Department of Sports Bioscience, School of Sports Sciences, Yoga and Education, Central University of Rajasthan, Ajmer, Rajasthan, India
e-mail: Hemanth.naick@curaj.ac.in

A. D. Singh et al. (eds.), *Sports Ergonomics*, Design Science and Innovation,
https://doi.org/10.1007/978-981-97-7804-1_9

corticosteroid medication, laser therapy, platelet therapy, glycerol patches, shock-wave therapy, exercise therapy, anti-inflammatory medications and surgical tendon repair to treat tendinosis, but the effective and standard treatment option to treat tendinopathies is still unavailable till date (Lopez and Jung 2015; Burton 2022). Hence, development of new methods and stratifies to treat tendinopathy is required in order to recover and rehabilitate the athletes in time to completion.

Exercise is highly recommended treatment option, in many musculoskeletal disorders along with medication in order to achieve strengthening and enhancing the functioning of the affected tendons (Luan et al. 2019; Dimitrios 2015). However, the existing training principles will increase the load on the tendon with no scientific evidence, and non-effective (Brumitt and Cuddeford 2015). Many studies have proved that physical therapy treatments and modalities will greatly help in speedy recovery of tendinopathy (Sussmilch-Leitch et al. 2012; Munari et al. 2022; Challoumas et al. 2019). Hence, it is essential for us to search and optimize a training strategy to support the athlete in speedy recovery and rehabilitation. To address this, in this study, we have formulated a new primary strategy by following scientific training principles and variables. We have designed a 4-week stretching exercise regimen that comprises the exercises straight leg rise, hamstring curl, prone quadriceps stretch and supine hamstring stretch in addition to the protection, rest, ice, compression and elevation (PRICE) protocol as described by Cormac Norton et al. (2016).

2 Materials and Methods

2.1 Subject Selection

This study was carried out on a 23-year-old state level football player, who was suffering from quadriceps tendinosis (QT). The subject presented localized pain and limitations in the movement near knee on left side. The subject had no other co-morbidities.

2.2 Study Design

The subject was diagnosed and reported to have quadriceps tendinosis in imaging (MRI) with the consultation of physician. Following, with the physician advice, we have designed a 4-week stretching exercise regimen along with PRICE protocol as described by Cormac Norton et al. (2016). Tests for leg strength, balance and biomarkers (Creatine kinase) for muscle injury were carried out every week. Post-intervention, diagnosis imaging test is carried out for quadriceps tendinosis.

2.3 *Training Protocol*

A 4-week basic exercise stretching protocol was followed after a general warm up and cools down session. The subject followed this regimen in 6 days in a week. The structured framework of workout session is as follows. As mentioned in Fig. 1.

Name of the regime

1. Straight leg rise: 10 repetitions / 2 sets
2. Hamstring curl: 10 repetitions/ 3 sets
3. Prone quadriceps stretch: 4 repetitions/ 2 sets
4. Supine hamstring stretch: 4 repetitions/2 sets

Fig. 1 Training protocol

2.4 Test Paradigm

As described above, the following test paradigm was formulated to test the leg strength, balance and biomarkers for muscle injury in serum.

(A) **Leg Muscle Strength Evaluation**

Static wall squat test: The subject was instructed to stand with his back to the wall and his knees and hips at a 90-degree angle. The subject was instructed to remain stationary for as long as is feasible. Duration of the time he hold the same position was calculated from the moment the stopwatch started until the subject's stance changed (Biscarini et al. 2022).

Squat test: The subject was asked to do squats as fast as possible in 30 s. The number of proper squats was counted (Hall et al. 2015).

(B) **Balance Test**

Static one-legged balance test: The individual was instructed to place his other leg on the injured leg and maintain pressure there and maintain the balance. The task was to continue for one minute with the subject's eyes covered. We observed the error that means loss of balance and recorded it (Matsuda et al. 2010).

The Four-square step test: In a square box, the subject was instructed to move as quickly as they could from 1 to 4 in a clockwise direction and then from 4 to 1 in an anticlockwise direction. The test was run three times, and each time, the time was recorded. It is used to assess the dynamic balance ability of the subject (Jee et al. 2013).

(C) **Muscle Injury Biomarker Test**

Serum Creatine Kinase (CK) Measurement: Serum creatine kinase was measured in subject once a week to monitor the recovery of the injured muscles (Baird et al. 2012a).

2.5 Statistical Analysis

The mean and standard deviation are used to present the data. As a single case study, the data points are very less so Mann–Whitney U test was used to check the significance at $P < 0.05$.

3 Result

3.1 Subject Details

The basic anthropometric data of the subject is given in Table 1.

The calculated BMI of the subject is 24.97 kg/m^2.

Table 1 Basic details of the subject

Subject (n)	Age (Years)	Height (cm)	Weight (kg)	BMI (kg/m^2)
1	23	165	68	24.97

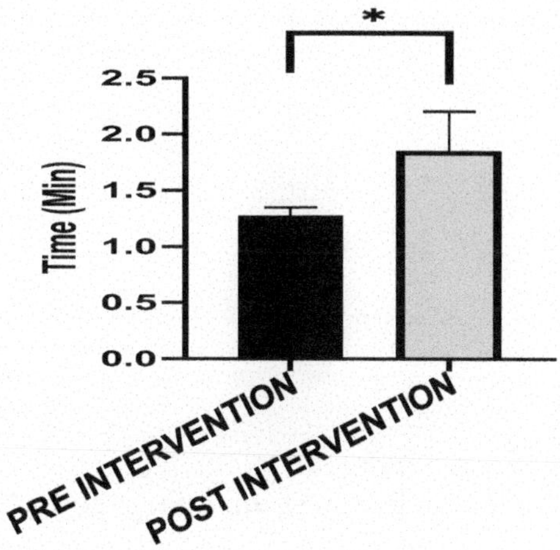

Fig. 2 Static squat test pre- and post-intervention

3.2 *Static Squat Test*

In static squat test, the quadriceps muscle plays a crucial role (Bush et al. 2015). Total time taken to do static squat in pre-intervention and post-intervention is calculated as described in methods along with every week of the intervention. In our study, we found that in Fig. 2, there is an increase in time between pre- and post-interventions, which clearly indicates that the muscle strength increases in subject after the 4 weeks of training protocol. We used Mann–Whitney U test and found the U-value is 1. The critical value of U at $p < 0.05$ is 2. Therefore, the result is significant at $P < 0.05$, the Z-score is 2.29783 and the p-value is 0,02,144. The result is significant at $P < 0.05$.

3.3 *Static One-Legged Balance Test*

The stability and balance capability of the subject is assessed as described in methods under static one-legged balance test. The results have showed that, no significant difference between pre- and post-interventions was observed as shown in Fig. 3. The U-value is 8.5. The critical value at $p < 0.05$ is 2. The reported Z-score is 0.73113 and the p-value is 0.4654 which is not significant $P > 0.05$.

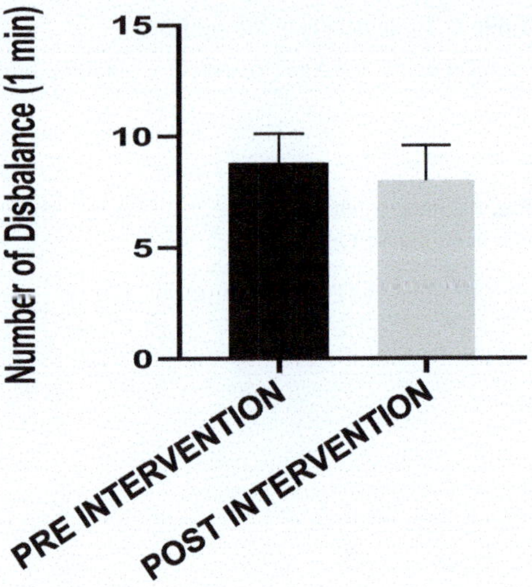

Fig. 3 Static one-legged balance test pre- and post-interventions

3.4 *The Four-Square Step Test (FSST)*

The dynamic balance of the subject was accessed as described in the methods section under four-square step test (FSST). In fast moving from one place to another quadriceps, muscle plays an important role (Alghadir and Anwer 2016). The results have shown that time take to complete the test is significantly reduced compared in post-intervention than the pre-intervention as shown in Fig. 4. The U-value is zero. The critical value of U at $p < 0.05$ is 2. Therefore, the result is significant at $p < 0.05$, where the Z-score is 2.50672 and the p-value is 0.01208 so the result is significant at $p < 0.05$.

3.5 *Squat Test*

Figure 5 shows the values recorded pre- and post-interventions, where the result shows that there is a significant increase in the number of squats performed per 30 s after interventions. This implies there is an increase in strength of the lower body and implies the improvement in muscle strength (Reggiani and Schiaffino 2020). The U-value for this test is 2. We know that the critical value of U at $p < 0.05$ is 2. Therefore, the result is significant at $p < 0.05$. The Z-score is -2.08893 along with the p-value is 0.03662 and the result is significant at $p < 0.05$.

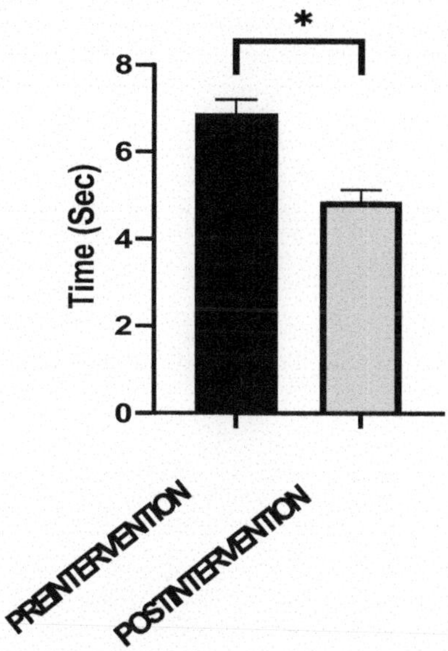

Fig. 4 Four-square step test pre- and post-interventions

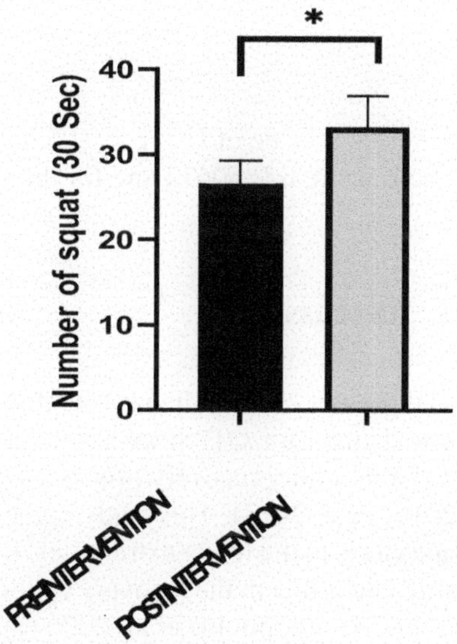

Fig. 5 Squat test (30 s) pre- and post-interventions

3.6 Serum Creatine Kinase Test:

In Fig. 6, there is a significant decrease in the serum CK level pre- and post-interventions. The decrease of CK level is a potent indicator of muscle recovery as well as speaks about tendon recovery (Baird et al. 2012b). The U-value is 0 where

Fig. 6 Serum creatine kinase level pre- and post-intervention

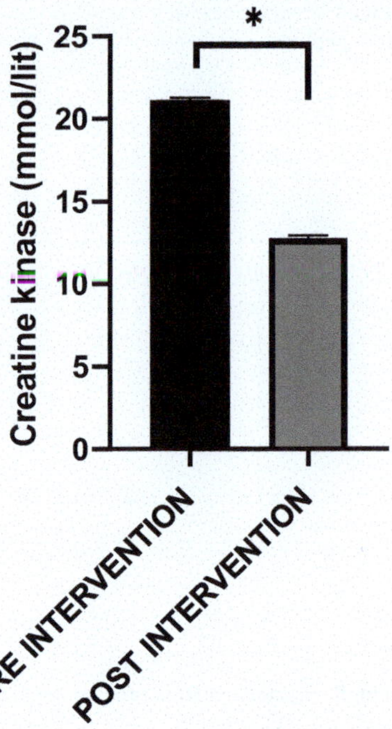

the critical value of U at $p < 0.05$ is 2. Therefore, the result is significant at $P < 0.05$. The Z-score is 2.50672 and for the same p-value is 0.01208.

4 Discussion

The present case study has established a 4-week stretching exercise as a rehabilitation therapy for QT. Since many studies has focused on advantages of stretching in different injury recovery processes (Sussmilch-Leitch et al. 2012; Challoumas et al. 2019; Page 2012). This study contributes for players suffering from sports injuries and stress in the lower extremities. Being an effective method, a specific stretching is still a dilemma in the recovery applications due to diversity in injuries (Babault et al. 2021). Stasinopoulos et al. 2006 (Stasinopoulos and Stasinopoulos 2016) observed exercise programme is more effective than, the other therapies such as ultrasound and transverse friction. Similar to the findings, study done by Manias et al. (2006) reported stretching as pain reducing formula for lateral elbow tendinopathy. Our study also observed a significant decline in pain due to QT at the end of the intervention. Hence, exercise performed under the guidance of professionals can significantly reduce the pain and improves the functionality of the injured muscles (Dimitrios and Pantelis 2013). Muscles stretching increases the elasticity and reduces the susceptibility from further damage (Maciejewska-Skrendo et al. 2020). A prospective follow-up study across 10 countries showed that reinjury risk is higher and very common in short recovery period (Gajhede-Knudsen et al. 2013). It is hypothesized that enhancing flexibility, as discovered in rabbit muscle resist the muscle for rupture by increasing the muscle length (Safran et al. 1988). Strength enhancement is also icing on the reducing the reinjury chances, as seen in our study, the static squat test for muscle strength flexibility and range of motion was improved. Dynamic balance improvement is linked with the lower extremity muscle strength and is a major research focus area nowadays (Kligyte et al. 2003; Leng et al. 2022). In the present case study, we have also reported that the stretching improves the balance and motor function. Proprioceptive, vestibular and visual information play key role in the postural adaptation in static and dynamic balance (Mickle et al. 2011).Therefore, we could not observe significant change in single leg static balance test. A yearlong varied flexibility and balancing exercise programme has also improved lower limb strength in the study of Taguchi et al. (2010).

Strength and balance improvement has been physical examination for the recovery process after performing eccentric, stretching and ultrasonic therapy (Dimitrios 2015). Clinical examination followed by injury biomarkers is equally important in the overall eradication claim for injury (Mukhopadhyay et al. 2011). Creatine kinase a protein, indirect marker for muscle damage (Baird et al. 2012b). Lopes et al. 2019 denoted that, CK and LDH were elevated 24–72 h after occurrence of simulated fight (Santos Silva Lopes et al. 2019). The present case reported significant decline in the parameter, which validation is required in large group cohort studies. Reduction in the CK level is been the strength of the recovery process since it reflects the improved muscle soreness and damage of the quadricep tendinosis. Our study limits in the overall rehabilitation followed by nutritional intervention. Based on the findings, implication of follow-up studies including larger group on mobility, mechanism and mode of action should be designed. Research focusing on development of easy to adapt, cost effective and absolute recovery process is needed.

5 Conclusion

In this study, we conclude that a 4-week stretching regimen will aid in the healing of musculotendinous injuries. We observed that the subject's lower limb muscle strength and balance levels were improved and creatine kinase levels were decreased in post- than pre-intervention. This suggests that, incorporation of stretching exercises along with PRICE will effectively help in recovering the tendinosis and greatly help in athletes completion performance.

Acknowledgements We acknowledge Central university of Rajasthan for supporting this work.

Conflict of Interest: The authors report no conflicts of interest and the authors report no competing financial interest.

Human ethics: The study was recommended by Institutional ethical committee.

References

Franchi M, Trirè A, Quaranta M, Orsini E, Ottani V (2007) Collagen structure of tendon relates to function. Sci World J 7:404–420. https://doi.org/10.1100/TSW.2007.92

Wu F, Nerlich M, Docheva D (2017) Tendon injuries: basic science and new repair proposals. EFORT Open Rev 2(7):332. https://doi.org/10.1302/2058-5241.2.160075

Abat F et al (2017) Current trends in tendinopathy: consensus of the ESSKA basic science committee. Part I: biology, biomechanics, anatomy and an exercise- based approach. J Exp Orthop 4(1):1–11. https://doi.org/10.1186/S40634-017-0092-6/METRICS

Andres BM, Murrell GAC (2008) Treatment of tendinopathy: what works, what does not, and what is on the horizon. Clin Orthop Relat Res 466(7):1539. https://doi.org/10.1007/S11999-008-0260-1

Fredberg U, Stengaard-Pedersen K (2008) Chronic tendinopathy tissue pathology, pain mechanisms, and etiology with a special focus on inflammation. Scand J Med Sci Sports 18(1):3–15. https://doi.org/10.1111/J.1600-0838.2007.00746.X

Charnoff J, Ponnarasu S, Naqvi U (2022) Tendinosis. Sport Heal A Multidiscip Approach 1(4):284–292. https://doi.org/10.1177/1941738109337778

Williamson PM et al (2021) Tendinopathy and tendon material response to load: what can we learn from small animal studies. Acta Biomater 134:43. https://doi.org/10.1016/J.ACTBIO.2021.07.046

Steinmann S, Pfeifer CG, Brochhausen C, Docheva D (2020) Spectrum of tendon pathologies: triggers, trails and end-state. Int J Mol Sci 21(3). https://doi.org/10.3390/IJMS21030844

Aicale R, Tarantino D, Maffulli N (2018) Overuse injuries in sport: a comprehensive overview. J Orthop Surg Res 13(1). https://doi.org/10.1186/S13018-018-1017-5

Kaux JF, Forthomme B, le Goff C, Crielaard JM, Croisier JL (2022) Current opinions on tendinopathy. J Sports Sci Med 10(2):238. Accessed: 28 Sep 2022 [Online]. Available: /pmc/articles/PMC3761855/

Ackermann PW, Renström P (2012) Tendinopathy in Sport. Sports Health 4(3):193. https://doi.org/10.1177/1941738112440957

Canosa-Carro L et al (2022) Current understanding of the diagnosis and management of the tendinopathy: an update from the lab to the clinical practice. Dis Mon 68(10):101314. https://doi.org/10.1016/J.DISAMONTH.2021.101314

Weinreb JH et al (2014) Tendon structure, disease, and imaging. Muscles Ligaments Tendons J 4(1):66. https://doi.org/10.32098/mltj.01.2014.12

El-Deek AMF, Dawood EMAEHH, Mohammed AAM (2019) Role of ultrasound versus magnetic resonance imaging in evaluation of non- osseous disorders causing wrist pain. Egypt J Radiol Nucl Med 50(1):1–7. https://doi.org/10.1186/S43055-019-0008-9/TABLES/5

Lopez RGL, Jung HG (2015) Achilles tendinosis: treatment options. Clin Orthop Surg 7(1):1. https://doi.org/10.4055/CIOS.2015.7.1.1

Burton I (2022) Combined extracorporeal shockwave therapy and exercise for the treatment of tendinopathy: a narrative review. Sport Med Heal Sci 4(1):8–17. https://doi.org/10.1016/J.SMHS.2021.11.002

Luan X et al (2019) Exercise as a prescription for patients with various diseases. J Sport Heal Sci 8(5):422. https://doi.org/10.1016/J.JSHS.2019.04.002

Dimitrios S (2015) Exercise for tendinopathy. World J Methodol 5(2):51. https://doi.org/10.5662/WJM.V5.I2.51

Brumitt J, Cuddeford T (2022) Current concepts of muscle and tendon adaptation to strength and conditioning. Int J Sports Phys Ther 10(6):748. Accessed: 28 Sep 2022 [Online]. Available: /pmc/articles/PMC4637912/

Sussmilch-Leitch SP, Collins NJ, Bialocerkowski AE, Warden SJ, Crossley KM (2012) Physical therapies for Achilles tendinopathy: systematic review and meta-analysis. J Foot Ankle Res 5(1):1–16. https://doi.org/10.1186/1757-1146-5-15/FIGURES/4

Munari D et al (2022) Effects of deep heating modalities on the morphological and elastic properties of the non-insertional region of achilles tendon: a pilot study. Int J Hyperth 39(1):222–228. https://doi.org/10.1080/02656736.2022.2026497

Challoumas D, Clifford C, Kirwan P, Millar NL (2019) How does surgery compare to sham surgery or physiotherapy as a treatment for tendinopathy? A systematic review of randomised trials. BMJ Open Sport Exerc Med 5(1):e000528. https://doi.org/10.1136/BMJSEM-2019-000528

Norton C (2016) How to use PRICE treatment for soft tissue injuries. Nurs Stand 30(52):48–52. https://doi.org/10.7748/NS.2016.E10506

Biscarini A, Dieni CV, Losavio R, Panichi R, Contemori S (2022) Effect of forefoot–rearfoot and pelvic–scapular weight shifts on lower-limb and lumbar muscle activity during static wall-squat exercises. Appl Sci 12(8):4037. https://doi.org/10.3390/APP12084037

Hall MP, Paik RS, Ware AJ, Mohr KJ, Limpisvasti O (2015) Neuromuscular evaluation with single-leg squat test at 6 months after anterior cruciate ligament reconstruction. Orthop J Sport Med 3(3). https://doi.org/10.1177/2325967115575900

Matsuda S, Demura S, Nagasawa Y (2010) Static one-legged balance in soccer players during use of a lifted leg. Percept Mot Skills 111(1):167–177. https://doi.org/10.2466/05.23.26.27.PMS.111.4.167-177

Jee H, Park J, Oh JG, Lee YH, Shin KA, Kim YJ (2013) Effect of a prolonged endurance marathon on vascular endothelial and inflammation markers in runners with exercise-induced hypertension. Am J Phys Med Rehabil 92(6):513–522. https://doi.org/10.1097/PHM.0B013E31829232DB

Baird MF, Graham SM, Baker JS, Bickerstaff GF (2012) Creatine- kinase- and exercise-related muscle damage implications for muscle performance and recovery. J Nutr Metab 2012. https://doi.org/10.1155/2012/960363

Bush JA, Blog GL, Kang J, Faigenbaum AD, Ratamess NA (2015) Effects of quadriceps strength after static and dynamic whole-body vibration exercise. J Strength Cond Res 29(5):1367–1377. https://doi.org/10.1519/JSC.0000000000000709

Alghadir A, Anwer S (2016) Effect of retro and forward walking on quadriceps muscle strength, pain, function, and mobility in patients with knee osteoarthritis: a protocol for a randomized controlled trial. BMC Musculoskelet Disord 17(1). https://doi.org/10.1186/S12891-016-1021-Z

Reggiani C, Schiaffino S (2020) Muscle hypertrophy and muscle strength: dependent or independent variables? A provocative review. Eur J Transl Myol 30(3):9311. https://doi.org/10.4081/EJTM.2020.9311

Baird MF, Graham SM, Baker JS, Bickerstaff GF (2012b) Creatine- kinase- and exercise-related muscle damage implications for muscle performance and recovery. J Nutr Metab 2012:13. https://doi.org/10.1155/2012/960363

Page P (2022) Current concepts in muscle stretching for exercise and rehabilitation. Int J Sports Phys Ther 7(1):109. Accessed: 28 Sep 2022 [Online]. Available: /pmc/articles/PMC3273886/

Babault N, Rodot G, Champelovier M, Cometti C (2021) A survey on stretching practices in women and men from various sports or physical activity programs. Int J Environ Res Public Health 18(8). https://doi.org/10.3390/IJERPH18083928/S1

Stasinopoulos D, Stasinopoulos I (2016) Comparison of effects of exercise programme, pulsed ultrasound and transverse friction in the treatment of chronic patellar tendinopathy. Clin Rehabil 18(4):347–352. https://doi.org/10.1191/0269215504CR757OA

Manias P, Stasinopoulos D (2006) A controlled clinical pilot trial to study the effectiveness of ice as a supplement to the exercise programme for the management of lateral elbow tendinopathy. Br J Sports Med 40(1):81–85. https://doi.org/10.1136/BJSM.2005.020909

Dimitrios S, Pantelis M (2013) Comparing two exercise programmes for the management of lateral elbow tendinopathy (Tennis Elbow/Lateral Epicondylitis)—a controlled clinical trial. Open Access J Sci Technol 1. https://doi.org/10.11131/2013/100013

Maciejewska-Skrendo A et al (2020) Genetics of muscle stiffness, muscle elasticity and explosive strength. J Hum Kinet 74(1):143. https://doi.org/10.2478/HUKIN-2020-0027

Gajhede-Knudsen M, Ekstrand J, Magnusson H, Maffulli N (2013) Recurrence of Achilles tendon injuries in elite male football players is more common after early return to play: an 11-year follow-up of the UEFA Champions League injury study. Br J Sports Med 47(12):763–768. https://doi.org/10.1136/BJSPORTS-2013-092271

Safran MR, Garrett WE, Seaber AV, Glisson RR, Ribbeck BM (1988) The role of warmup in muscular injury prevention. Am J Sports Med 16(2):123–129. https://doi.org/10.1177/036354658801600206

Kligyte I, Lundy-Ekman L, Medeiros JM (2022) Relationship between lower extremity muscle strength and dynamic balance in people post-stroke. Medicina (Kaunas) 39(2):122–128. Accessed: 14 Sep 2022 [Online]. Available: https://europepmc.org/article/med/12626864

Leng H et al (2022) Correlation between dynamic balance ability and lower limb muscle strength of university students. World J Adv Res Rev 14(3):423–430. https://doi.org/10.30574/WJARR.2022.14.3.0546. https://wjarr.com/sites/default/files/WJARR-2022-0546.pdf

Mickle KJ, Munro BJ, Steele JR (2011) Gender and age affect balance performance in primary school-aged children. J Sci Med Sport 14(3):243–248. https://doi.org/10.1016/J.JSAMS.2010.11.002

Taguchi N, Higaki Y, Inoue S, Kimura H, Tanaka K (2010) Effects of a 12-month multicomponent exercise program on physical performance, daily physical activity, and quality of life in very elderly people with minor disabilities: an intervention study. J Epidemiol 20(1):21–29. https://doi.org/10.2188/JEA.JE20081033

Mukhopadhyay M, Sinha R, Pal M, Bhattacharyya S, Dan A, Roy MM (2011) Role of common biochemical markers for the assessment of fracture union. Indian J Clin Biochem 26(3):274–278. https://doi.org/10.1007/S12291-011-0143-1/FIGURES/3

Santos Silva Lopes J, Monteiro de Magalhães Neto A, Oliveira Gonçalves LC, Lourenço Alves PR, Castilho de Almeida A, Marlise Balbinotti Andrade C (2019) Kinetics of muscle damage biomarkers at moments subsequent to a fight in Brazilian Jiu-Jitsu practice by disabled athletes. Front Physiol 10:1055. https://doi.org/10.3389/FPHYS.2019.01055/BIBTEX

Chapter 9
Physiological Responses During Endurance Speed Training of Field Hockey Athletes

Ekta Kapri, Manju Mehta, and Kiran Singh

1 Introduction

Field hockey is one of the most popular and widely performed sports in the world. As an intermittent activity, it places substantial demands on players' aerobic and anaerobic systems. Athletes are required to perform repeated bursts of sprinting, rapid directional changes, and sustained physical effort over extended periods. These physiological demands necessitate an optimal anthropometric profile, which enables players to cope with the high physical workload inherent (Montgomery 2006; Norton et al. 1996) that is optimal for the high physiological workload that the game places on its players (Reilly and Borrie, 1992; Elferink-Gemser et al. 2006).

Maximum workload in the field activity generally leads to maximum physical exertion. Activity load demand on the athletes varies according to training and competitive levels, which reflects the interplay between internal and external load on the athletes. These activity loads on the athletes during and after practice either affect the physical and physiological characteristics or get affected by these parameters. It is critical to assess the present physical demands and workload on the players during high-level competition in order to establish conditioning methods to enhance aerobic and anaerobic performance of team sport players. Heart rate (HR) and blood lactate (bLA) concentration are two key physiological markers frequently used in sports science to evaluate an athlete's aerobic capacity and response to endurance exercises (Janssen 2001). During exercise or any field activity, changes in heart rate

E. Kapri (✉) · M. Mehta · K. Singh
Department of Family Resource Management, CCSHAU, Hisar, Haryana, India
e-mail: kapriekta70@gmail.com

M. Mehta
e-mail: manjuijmehta19@gmail.com

K. Singh
e-mail: kiransingh.in@gmail.com

A. D. Singh et al. (eds.), *Sports Ergonomics*, Design Science and Innovation,
https://doi.org/10.1007/978-981-97-7804-1_10

(HR) and heart beat volume are modified in response to the body's growing metabolic demand (Ismail et al. 2013). Resting HR between 60 and 80 bpm can rise up to about 200 with exercise and can be returned to normal within 1–2 h (Tamer 1991). These variations in heart rate provide valuable insights into an athlete's cardiovascular fitness and capacity to sustain physical exertion over time. In addition to heart rate, blood lactate concentration serves as an important indicator of an athlete's ability to produce and maintain energy during high-intensity activity. Lactate levels typically range from 5-10 mg or 0.5–1 mmol L^{-1} values in 100 cc blood, under normal conditions (Ferry et al. 1988), and for trained individuals during rest, lactic acid level is 0.5–2.2 mmol L^{-1} ($\cong$ 4.5–19.8 mg/dl) (Hermansen and Stensvold 1972). Lactate is the sole metabolic indicator that shows the ability of muscles for athletic performance for keeping sustainable and balanced energy required by the activity. As a result, during the exercise, the athlete creates the most quantity of energy per time (Ismail et al. 2013). Individual times for attaining the maximum lactate level may depend on the exercise time, frequency, and duration, exercise force, and physical fitness of the individual (Fox et al. 1988). When compared to trained persons, lactic acid rise is more of a reaction to exerting the same effort (Ural 1972). The primary objective of this study is to investigate the heart rate and blood lactate responses of field hockey players during endurance speed training. By examining these physiological responses, the study aims to provide deeper insights into how field hockey players manage the physical demands of the game and recover from sustained physical exertion. Additionally, this study opens up opportunities to explore the effects of varying training intensities on recovery times and performance outcomes across different playing positions.

2 Method

The main objective of the present study aimed to investigate the heart rate and blood lactate response of field hockey players during endurance speed training.

2.1 *Sampling Procedure*

Female field hockey team practicing under sports training center (STC), SAI, Giri center, comprises around 40–45 athletes in total, summing up all age groups: sub-junior (under 17), junior (under 20), and senior (20–above). Thirty subjects were purposively selected on the basis of consistency in performance and physical fitness. The physical fitness of subjects was ascertained by physical fitness test along with physical and physiological parameters. This study procedure was approved by the university ethical committee. In charge of the sports training center (SAI) and coaches of female field hockey consented the study and allowed to approach the players to participate in the study. Athletes signed a consent form and required information had been imparted to them. Each procedure was explained to the individual athlete verbally and whenever required. Athletes were introduced to the test and procedure prior to data collection.

2.2 *Subjects*

There are eleven players on the ground of hockey—ten field players and one goalkeeper. Players filed position is in the ratio of 1:2:4:4 that is 1 goalkeeper, 2 defenders (full back), 4 center half (midfielders), and 4 forward players. In the present study, out of thirty subjects, 40.00% of the subjects' field position was center half, 30.00% was forward players, 16.67% was defenders, and 13.33% was goalkeepers.

Thirty young female field hockey athletes of national level gave their written informed consent to participate in this study. Their training program during the three months prior the study represented approximately 2.5 ± 0.5 h daily morning and evening training including warm up, training, and warm down sessions. Goal keeper and full back players had competitive experience of 3.75 ± 2.99 and 3.20 ± 2.59 years, whereas center half and forward players had 4.63 ± 1.76 and 4.11 ± 2.36 years of competitive experience.

2.3 *Inclusion Criteria*

Physical fitness test give idea about aerobic capacity, endurance, and energy efficiency and work productivity of the athletes. In the present study, recovery level 1 (Yo-Yo IR1) was selected to record the fitness level of the subjects. This is the test of aerobic capacity (physical fitness), conceived around the early 1990s by Jens Bangsbo (Bangsbo and Lindquist 1992) a Danish soccer physiologist, and then described in a 2008 paper, "The Yo-Yo Intermittent Recovery Test" (Bangsbo et al. 2008) (Table 1).

Table 1 Assessment of physical fitness of the subjects through Yo-Yo IR1

Yo-Yo IR1	Mean ± SD	Min	Max
Test score	16.1 ± 1.2	13.20	18.10
Number of shuttle	28.00 ± 09.00	09.00	44.00
Cumulative no. of shuttle	57.00 ± 19.00	18.00	88.00
Total distance (m)	1136.0 ± 370.70	360.00	1760.0
Running speed (km/h)	15.40 ± 0.60	14.00	16.50
Time (min:s)	09:24 ± 03:01	03:01	14:24

Final 30 subjects were selected on the basis of the physical fitness level determine with Yo-Yo intermittent test level The subject's fitness was evaluated based on the test score; the higher the test score, the greater the physical fitness. Subjects with the test score above 13 were selected for the study. The minimum test score was 13.20, which was obtained by covering 360 m total distance in 3 min 01 s with 14 km/h running speed. Average test score of 16.1 ± 1.2 covered 1136.0 ± 370.70 m of total distance in 9 min 24 s with 15.40 ± 0.60 km/h of running speed. Maximum test score was 18.1 obtained in 14 min 24 s, covering 1760 m distance (44 shuttles) with 16.5 km/h of running speed.

The Yo-Yo IR1test results revealed that with increase in the test score, number of shuttle, running speed, total distance covered by the subjects, total time taken (to get the test score), and VO_2 max has also increased. Table 1 further showed, in the test score 13.1 to 16.1, there was gradual increase in the running velocity and VO_2 max; but in test score 16.1 to 16.8, running velocity remains same, whereas there was gradual increase in the VO_2 max; similarly in test score 17.1 to 17.8, running velocity remains same, whereas there was gradual increase in the VO_2 max.

2.4 *Physical Measurements*

The obtained reactance and resistance values as well as data on subjects' age, height, weight, fat percentage, and BMI were measured. Body height of the subjects was determined to the nearest 1 mm using an anthropometric scale. The weight and BMI of the subjects were measured by BIA system (Omron body composition monitor, HBF-375 Karada scan) with the measurement error of 0.5 kg. BF % was calculated from Siri equation (Guerra et al. 2010).

$$\text{Fat}\% = (4.95/D - 4.5) \times 100$$

2.5 *Fatigue Assessment*

Endurance speed training was selected for the fatigue analysis, in which subjects performed continuous running for 25 min at 12 km/h speed. Variation was noticed in the speed of subjects during performing task.

2.6 Blood Lactate Sampling

The blood lactate concentration was measured through finger-stick (capillary) blood sampling. Briefly, 0.2 μL of blood was collected before being analyzed using an automated blood lactate analyzer (EKF Diagnostics Lactate Scout 4).

2.7 Selection of Training

In the morning session, the weekly routine was divided into speed, agility, strength, and endurance speed work. Each type of training has a different impact on the subjects' metabolic systems. In the speed, agility, and strength training, the session was divided into 3–4 sets with 1, 2, 3 or 5 min of pause or recovery (depending on the type of activity and subjects' performance), whereas in the endurance speed training, duration and intensity were pre-fixed by the coach in every training session. It might be 30 min of running with 8–10 km/h speed or 20 min of running with 10–12 km/h speed. In the present study, endurance speed training was selected for the fatigue analysis, in which subjects performed continuous running for 25 min at 12 km/h speed.

2.8 Lactate Test Protocol

Blood sample was collected in two phases: resting and after activity. A resting sample was taken before the session. The second phase of the lactic acid test was done immediate after activity, after 5 and 10 min, of active recovery. No blood sample was collected in between the activity. This LA detection method corresponded well with the planned sample collection protocol by Hermansen and Osnes (1972) and Ohkuwa et al. (2009).

2.9 Statistical Analysis

MS Excel for Windows 7 and SPSS were used to conduct the analyses. Descriptive statistics (Means SD) were used to present the data. A Pearson correlation coefficient was also tested to determine the relationship between heart rate and blood lactate levels (Lopes et al. 2012). Using significant values for Pearson's correlation coefficient, the alpha value was set at 0.01 and 0.05 to assess the relation between variables. The Shapiro–Wilk test was used to test data normality. Linear relation was calculated between HR and VO_2max and HR and bLA (mmol L^{-1}).

3 Result

3.1 Physical Characteristic of Field Hockey Players

Field hockey demands intermittent exercise due to multiple brief intense activities, like sprint, quick turns, and tackles. Therefore, the aerobic and anaerobic capacity of the field hockey athletes was evaluate with the physical fitness test. Understanding the level of fitness prior to the physical and physiological assessment helped to select the potential subject's for the study and also helped to design the test protocol for the study.

Table 2 depicts the physical characteristics including BMI, fat%, and Vo_2max of the players of different position. Age, weight, height, BMI, fat%, and VO_2max of the subjects were significantly similar.

Table 2 showed the descriptive statistics of the subjects. Subject's average age was 16.70 ± 02.17 years, height average is 162.17 ± 06.10 cm, and weight average is 51.82 ± 05.23 kg. The average BMI of the subjects was 19.70 ± 1.70 (kg/m^2) which falls under the overweight category (Pasco et al. 2014), with 19.24 ± 3.00% of body fat. As described by the American Council on exercise, body fat percentage is categorized into the following ranges: essential fat (2–5%), athletes (6–13%), fitness (14–17%), average (18–24%), and obese (above 25%) (Muth 2009). In line with this, in the present study, subjects' body fat percent was categorized as average. No positional difference was evident in any of the variable among the subjects. In line with this, similar study was reported by Ghosh et al. (1991).

3.2 Heart Rate of the Subjects

Table 3 explained that in the morning session, subjects mean HR was recorded 60.23 ± 5.41 bpm, immediate after activity mean HR of the subjects was increased by 171.39% (↑), whereas after 5 and 10 min of active recovery HR was gradually decreased by 30.64% (↓) to 13.44% (↓).

In the morning session, goal keepers resting heart rate was recorded highest (81.25 ± 6.40 bpm), which increased by 96.8% immediate after 25 min of endurance speed training and decreased up to 115.75 ± 07.63 bpm in 5 min, and around 13 bpm in 10 min of active recovery. On the other hand, fullback, center half, and forward players, HR rest was almost similar 76.20 ± 6.06 bpm, 73.00 ± 4.73 bpm 74.22 ± 5.78 bpm, respectively, which was increased by 104.8%, 122.22%, and 130.21% immediate after activity and decreased by 50.12–14.60 bpm, 48.89–16.83 bpm, and 56.45–14.66 bpm in 5–10 min of active recovery.

Table 2 Physical characteristic of field hockey players

Field position	Goal keeper (GK)	Full back (FB)	Center half (CH)	Forward (F)	Mean ± SD	*p*-value
Age (years)	17.50 ± 2.38	16.00 ± 1.00	16.67 ± 2.10	16.78 ± 2.77	16.70 ± 02.17	0.019*
Weight (kg)	55.20 ± 3.53	52.84 ± 4.85	52.21 ± 6.19	49.22 ± 3.97	51.82 ± 05.23	0.87
Height (cm)	164.47 ± 5.76	161.64 ± 0.64	162.42 ± 6.13	161.11 ± 8.13	162.17 ± 06.10	0.37
BMI (kg/m^2)	20.45 ± 1.75	20.22 ± 1.85	19.77 ± 1.94	18.99 ± 1.20	19.70 ± 1.70	0.31
Fat %	19.46 ± 3.92	19.50 ± 2.81	19.63 ± 2.90	18.48 ± 3.23	19.24 ± 3.00	0.18
Vo_2max (ml/min/kg)	44.63 ± 4.66	46.08 ± 2.69	45.47 ± 3.11	47.08 ± 2.73	45.87 ± 05.82	0.05*

* p-value < 0.05

Table 3 Heart rate of the subjects according to field position

Field position	Resting	Immediate after	R1 (5 min)	R2 (10 min)
Goal keeper (GK)	81.25 ± 6.40	159.92 ± 12.08	115.75 ± 07.63	103.25 ± 04.43
Full back (FB)	76.20 ± 6.06	156.07 ± 11.21	109.80 ± 02.95	95.20 ± 04.44
Center half (CH)	73.00 ± 4.73	162.14 ± 17.19	113.25 ± 18.67	96.42 ± 11.80
Forward (F)	74.22 ± 5.78	170.89 + 20.89	114.44 ± 11.40	99.78 ± 01.89
Mean ± SD	60.23 ± 5.41	163.46 ± 17.14	113.37 ± 13.37	98.13 ± 8.45

3.3 *Blood Lactate Concentration of the Subjects*

Subjects practice in different positions having different level of physical fitness affects the performance level of the subjects and depends upon the various physical (anthropometry and body composition) and physiological (Vo_2max and heart rate) factors. Another important factor which affects the performance of the subjects is metabolites, which can be the byproduct of aerobic and anaerobic metabolism.

Table 4 explained the change in blood lactate concentration of subjects according to their field position. Blood lactate level of the subjects was measured in the morning session, before and after endurance speed practice. Mean resting lactate level was 2.45 mmol L^{-1} with 0.91 mmol L^{-1} of standard deviation, which increased by which 252.24% (↑) immediate after 25 min of continuous running with 12 km/h speed; whereas after 5 min of active recovery blood lactate level decreased by 18.31% (↓) and up to 10 min of recovery lactate level further decreased by 19.15%(↓).

For all four field positions, resting blood lactate level (mmol L^{-1}) was almost similar, i.e., 02.63 ± 00.78 mmol L^{-1} (GK), 02.42 ± 01.10 mmol L^{-1} (CH), 02.26 ± 00.92 mmol L^{-1} (FB), and 02.22 ± 00.80 mmol L^{-1} (F). Immediate after activity, inflation has seen in the blood lactate level (mmol L^{-1}) of all the subjects (as exhibits in Fig. 1).

Table 4 Blood lactate concentration of subjects as per field position

Field position	Resting	Immediate after	R1 (5 min)	R2 (10 min)
Goal keeper (GK)	02.63 ± 00.78	14.13 ± 04.56	06.95 ± 02.79	07.28 ± 05.99
Full back (FB)	02.26 ± 00.92	05.88 ± 06.48	05.78 ± 05.01	04.94 ± 04.44
Center half (CH)	02.42 ± 01.10	07.99 ± 06.34	07.57 ± 05.19	05.28 ± 04.02
Forward (F)	02.22 ± 00.80	08.56 ± 04.26	07.11 ± 04.64	05.97 ± 03.58
Mean ± SD	02.45 ± 00.91	08.63 ± 05.82	07.05 ± 04.56	05.70 ± 04.08

*R*1 and *R*2 replication

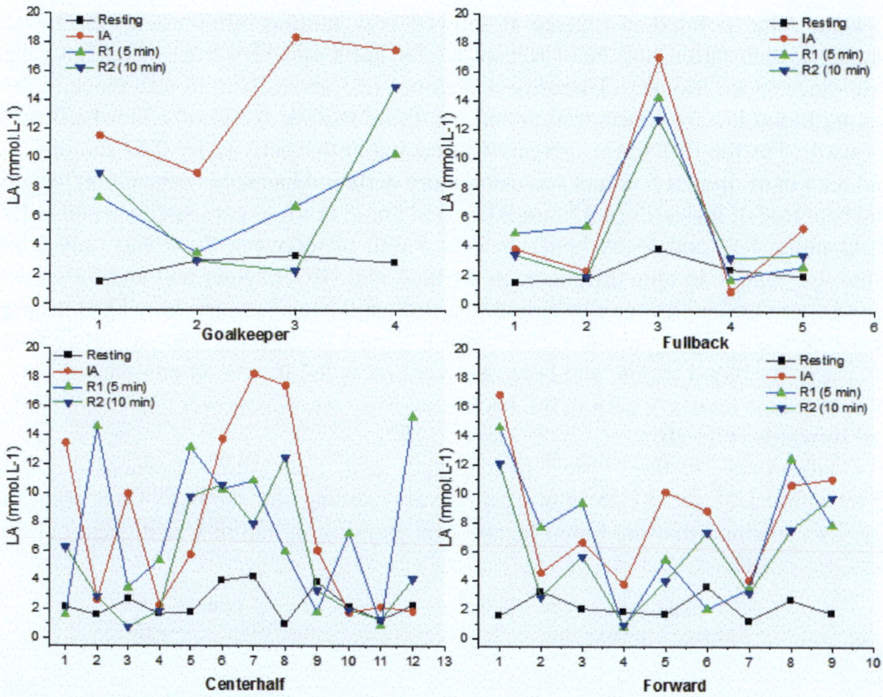

Fig. 1 Change in blood lactate level before and after activity (endurance speed)

Figure 1 concluded that the subjects' resting blood lactate (mmol L^{-1}) levels were almost identical. Blood lactate range was different for the subjects playing in different positions such as goalkeeper blood lactate ranges between 1.5 and 18.4 mmol L^{-1}; for center half, blood lactate range was 0.8–18.3 mmol L^{-1}; for full back, blood lactate range was 0.9–17.1 mmol L^{-1}; and for forwards, blood lactate range was 0.8–16.9 mmol L^{-1}. After 25 min of continuous running, lactate level in the blood was gradually raised. In recovery period, subjects blood lactate level can be increased or decreased that depends on the type and intensity of activity subjects had performed and type of recovery (active or passive) they are performing.

3.4 Relation Between Subjects' Heart Rate and Blood Lactate Concentration in the Morning Session (Endurance Speed)

Figure exhibits the graphical representation of heart rate and blood lactate concentration in the morning session during endurance speed training.

Data value pointed in Fig. 2c, d showed very minute difference in the blood lactate concentration and heart rate of center halfs and forwards, throughout the endurance speed training. There was 0.2 mmol L^{-1} and 1.22 bpm difference in the resting blood lactate concentration and heart rate (respectively) of center halfs and forwards, but the difference increased immediate after activity by 0.57 mmol L^{-1} and 8.75 bpm. In first 5 min of recovery, more deflation has seen in the blood lactate and hear rate of forwards (1.45 mmol L^{-1} and 56.45 bpm) as compare to center halfs (0.42 mmol L^{-1} and 48.89 bpm). In next 5 min of recovery, there was only 0.69 mmol L^{-1} and 3.36 bpm difference in the bLA and HR of center half and forwards.

As compared with center halfs and forwards, full backs' resting blood lactate and hear rate were slightly above (2.26 mmol L^{-1} and 76.2 bpm), whereas immediate after activity blood lactate and hear rate were recorded lowest among the subjects. With the same recovery period, full backs' recovery rate was slower than center halfs and forwards (Fig. 2b).

Goalkeepers' resting blood lactate and hear rate were measured highest (2.63 mmol L^{-1} and 81.25 bpm, respectively) among the subjects. Data point in Fig. 2a explained that the blood lactate concentration inflation was highest (11.50

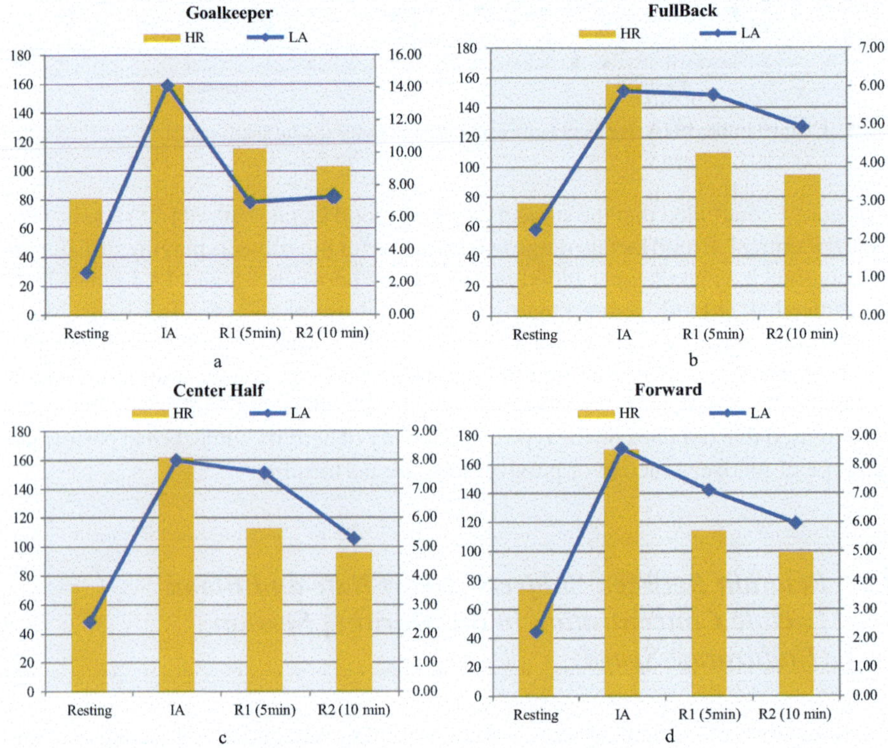

Fig. 2 *bLA* blood lactate concentration; *HR* heart rate

mmol L^{-1} point↑) but the heart rate was lower than center halfs and forwards. Furthermore, highest bLA drop was noted in first 5 min of recovery (7.18 mmol L^{-1}), but the HR recovery was slower (44.17 bpm), and in next 5 min of recovery, there was 0.33 mmol L^{-1} inflation which was recorded in blood lactate, but the heart rate gradually went down up to 10 min of recovery.

3.5 Relation Among Heart Rate and Blood Lactate Concentration

Table 5 explains the correlation ($\alpha = 0.05$) among the subjects' heart rate and blood lactate concentration in the morning training of endurance speed. Heart rate and blood lactate of the subjects were recorded before the activity (resting), immediate after the activity, in 5 min, and 10 min of active recovery. Resting bLA shows significantly ($\alpha = 0.05$) weak relation with blood lactate (IA) $r = 0.378$, whereas it had negligible relation with recovery blood lactate concentration [$R1$ ($r = 0.125$) and $R2$ ($r = 0.151$)]. With heart rate, resting bLA have negative to low positive relation (ranges – 0.282 to 0.462); but recovery HR showed positively weak ($r = 0.462$) relation with resting bLA. Blood lactate concentration immediate after (bLA IA) activity had weak correlation with R1LA ($r = 0.301$) while on the contrary it had significantly ($\alpha = 0.05$) strong positive relation with R2LA ($r = 0.715$).

With heart rate, bLA IA showed negative to moderate positive relation (ranges - 0.057 to 0.519); where with HRrest, bLA IA showed negative relation ($r = -0.057$), whereas with HR IA, it showed negligible relation ($r = 0.269$); with R1HR, weak

Table 5 Pearson correlation among heart rate and blood lactate concentration

Variables	LArest	LA IA	R1LA (5 min)	R2LA (10 min)	HRrest	HR IA	R1HR (5 min)	R2HR (10 min)
bLArest	1							
bLA IA	0.378	1						
R1LA (5 min)	0.125	0.301	1					
R2A (10 min)	0.151	0.715*	0.535*	1				
HRrest	– 0.282	– 0.057	0.107	0.035	1			
HR IA	0.036	0.269	0.131	0.054	0.066	1		
R1HR (5 min)	0.278	0.358	0.162	0.158	– 0.106	0.206	1	
R2HR (10 min)	0.462	0.519*	0.152	0.253	– 0.345	0.263	0.472	1

bLA blood lactate; *IA* immediate after; *R*1 and *R*2 replication; *HR* heart rate; *HRrest* resting heart rate; $\alpha = 0.05$ level of significance

relation ($r = 0.358$) was established, and R2 HR had moderately significant relation ($r = 0.519$). Recovery blood lactate concentration was moderately correlated with each other ($r = 0.535$), whereas both R1LA and R2LA showed very weak relation with heart rate (HRrest, HR IA, R1HR, and R2HR) which ranges 0.035–0.162. Heart rate before and after the training (HRrest, HR IA, R1HR, and R2HR) had negative to weak correlation among each other (ranges − 0.345 to 0.472). Table 5 concludes that accumulated blood lactate and elevated heart rate had significantly positive recovery but both showed unstable correlation throughout the session.

4 Discussion

The purpose of this study was to analyze the heart rate and blood lactate response of field hockey players during endurance speed training.

In any sport, a high workload leads to a high level of physical exertion and tiredness. It is determined by the athletes' physical and physiological characteristics, which all are statistically associated ($\alpha = 0.05$). As field hockey is very fast moving sports, it demands intermittent exercise to perform a task in the field with perfect human physique, body composition, respiratory and circulatory system, body metabolic process, and body force and strength (Elferink-Gemser et al. 2007; Calo et al. 2009; Ucan 2015). The level of fitness of the subjects was assessed through Yo-Yo intermittent recovery 1, which is the most extensively studied fitness tests in sports science (Bangsbo et al. 2008). Veerappa (2018) reported that Indian field hockey players' highest Y0-Y0 test score was 21.1 while the lowest was reported 17.7. On the other hand in the present study, the minimum test score was 13.20 and maximum test score was 18.1. The average test score was 16.1 ± 1.2, covered in 9 min 24 s with 15.40 ± 0.60 km/h of running speed.

4.1 Assessment of Physiological Parameter of Subjects

Mean age, weight, and height of the subjects are 16.70 ± 02.17 years, 51.82 ± 05.23 kg, and 162.17 ± 06.10 cm, respectively. Almost similar findings were reported by Ucan (2015) in a study on elite national-level field hockey players (age: 19.6 ± 0.8 years; weight: 56 ± 6.1 kgs; height: 162.1 ± 4.8 cm; playing experience: 6.5 ± 0.4 years). In the present study, weight and BMI were measured by bioelectrical impedance analysis (BIA) technique. In some early research, body composition researchers (Richard Boileau, Lary Golding, Tim Cohman, Karen Segal, Wayne Sinning, and Jack Wilmore) have developed newer technique and prediction equation with lower standard errors, comparable to skin fold technique. In the present study, mean weight 51.82 ± 05.23 (kg) with 19.70 ± 1.70 BMI (kg/m^2) and fat percentage (19.24 ± 3.00%) fall under normal category (Morgan 1973).

Along with the anthropometric parameters, field hockey training also affects the physiological parameters of the players, and the physiological parameters are directly related with energy systems. As hockey is physically demanding sports, it necessitates a lot of energy from each of the three systems. The aerobic system is important during prolonged intermittent exercise, and high-intensity efforts rely on the anaerobic energy systems, ATP-Cr (adenosine triphosphate phosphocreatine) for the intermediate, and anaerobic glycolysis for short term (Wood 2021). These parameters are often inter-correlated due to dependence of one on another and suggest adaptations related to training (Montgomery 2006).

Assessing the training load of the athletes' heart rate is the most frequent method, especially for the long duration sports (Lopes et al. 2012). Subject's minimum and maximum resting heart rate was recorded 51 bpm and 68 bpm, respectively (Sharma and Kailashiya, 2017). Subjects usually involved in strength and conditioning training focused on aerobic fitness in the morning session. Therefore, subjects HR was recorded higher in the morning session. In the team sports, player position on the field might affect the physical load and exertion on the player. The present study explained according to subjects' field position. Study by Ghosh et al. (1991) on field hockey players concluded that forward players' HR was highest (151.1 ± 19.5—171.6 ± 16.5), followed by center half (153.3 ± 16.9—167.1 ± 16.1), full back (146.4 ± 16.5—158.7 ± 17.4), and goal keeper (122.8 ± 8.4—129.2 ± 10.6).

As VO_2max depend on the air exchange in the lungs, heart rate, arterial blood flow to the muscles, and oxygen demand by the tissue (used by mitochondria) (Wagner 1996), it also affects the biochemical metabolism in the body. More than 90% of a game is performed by aerobic metabolism (Bangsbo 1994), and the average intensity is around the anaerobic lactate threshold (80–90% of maximal heart rate) (Bangsbo 1994; Helgerud et al. 2001). Over specific time period with certain level of intensity, the players either exercise above (accumulating lactate) or below (oxidizing the accumulated lactate) the threshold (Helgerud et al. 2001). Along with the physical and physiological factors, metabolite can be an important factor that affects the performance of the athletes.

Lactate buildup has generally been attributed to lower athletic performance, with many people believing in the lactate acidosis theory. Lactate buildup occurs because more pyruvate is converted to lactate by lactate dehydrogenase, principally as a result of changes in intramuscular redox state, and because oxidation of the excess lactate requires redistribution of the lactate via blood flow to other muscles and the heart and liver (Gladden 2004; Wasserman et al. 1986).

Before warm up variation in lactate level was 2.22–2.63 mmol L^{-1}. As per Ghosh et al. (1991), it was lowest for forward player and highest for goalkeepers. Immediate after 25 min of continuous running mean bLA was increased by 252.24% and ranged between 05.88 and 14.13 mmol L^{-1} (Reilly and Borrie 1992) (minimum for full back player and maximum for goalkeepers'), whereas after 5 min of active recovery bLA was decreased by 18.31% (range 05.78–07.11 mmol L^{-1}) (minimum for full back player and maximum for fullback), similarly up to 10 min of recovery bLA gradually decreased by 19.15% (ranged 04.94–07.28 mmol L^{-1}) (Ghosh et al. 1991) (minimum for full back player and maximum for goalkeepers').

Study further revealed that goalkeepers' bLA inflation and recovery rate was highest whereas full back players' bLA inflation and recovery rate was lowest. Minimum accumulation of bLA and higher recovery of bLA is good indicator for an athlete, in the context of sports performance. Study also reported the diversified recovery of the subjects' blood lactate concentration.

4.2 *Assessment of Physiological Parameter of Subjects*

Furthermore, workout intensity is frequently defined in relative terms when determining the ideal exercise intensities for each athlete (Mann et al. 2013). The percentage of maximum aerobic effect is typically used to measure exercise intensity (Meyer et al. 1999). HR may be used to achieve certain bLA (mmol L^{-1}) responses based on the relationship between HR and bLA (mmol L^{-1}) recorded during exercise testing, making HR a valuable tool for training guidance (Bramell 2016).

Data revealed the minute difference in the blood lactate concentration and heart rate of center halfs and forward players, throughout the endurance speed training. Fullback players' resting blood lactates and HRrest were slightly above whereas immediately after activity blood lactate and HR were recorded lowest among the subjects. With the same recovery period, fullback players' recovery rate was slower than center half and forward players. Goalkeepers' resting blood lactate and HR were highest, whereas highest LA drop was also noted in first 5 min of recovery, but slight inflation was recorded in blood lactate but the heart rate gradually went down up to 10 min of recovery. Mean blood lactate concentration before activity, during activity, and recovery period also showed week correlation with HRrest, HR IA, R1HR, and R2HR. Study further concluded that accumulated bLA and elevated HR had significantly positive recovery, but recovery bLA (R1LA and R2LA) showed very weak relation ($\alpha = 0.05$) with heart rate (HRrest, HR IA, R1HR, and R2HR) which ranges 0.035–0.162. Study showed that blood lactate, oxygen consumption, and heart rate has significant ($\alpha = 0.001$) linear relation with perceived exertion of the subjects.

The main outcomes of such testing are maximum values for oxygen consumption (VO_2max), heart rate (HR), and blood lactate concentration (bLa) as well as relationship between HR and VO_2max, HR and [bLa], [bLa] and VO_2max (Vergès et al. 2003). In the present study (Fig. 3), HR showed linear relation with VO_2max ($r = 0.910$) and bLA (mmol L^{-1}) ($r = 0.841$) ($\alpha = 0.05$).

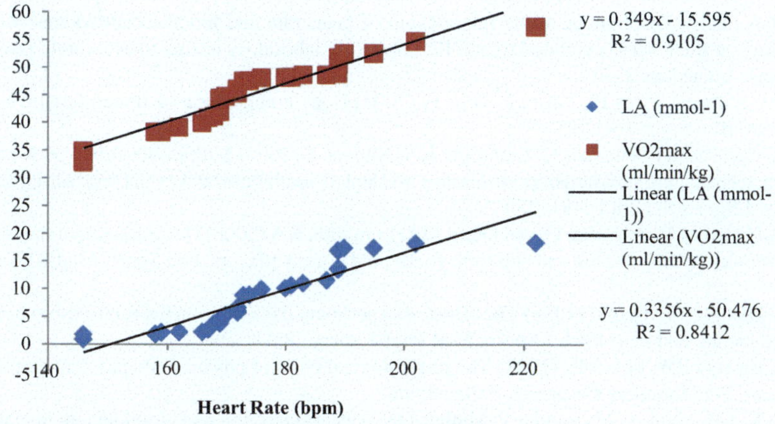

Fig. 3 Linear relation between HR-VO2max and HR-bLA

5 Conclusion

In the present study, three types of changes have been noticed in the blood lactate level after the activity. Up to 10 min of active recovery, (a) subjects lactate level increased in first 5 min and then dropped in next 5 min, (b) subjects' blood lactate level gradually decreases in 5 min and 10 min of recovery, and (c) subjects' blood lactate level drop in first 5 min but raise in next 5 min of recovery. Study further concluded that with increase in the heart rate, blood lactate concentration has also increased, whereas during recovery heart rate dropped gradually and faster than blood lactate. Among all the subjects, forwards showed good blood lactate recovery. The position-specific lactate responses and the strong correlations between physiological parameters provide valuable insights for coaches and trainers. Athletes with lower lactate accumulation and faster recovery rates may have a physiological advantage in sports performance. These findings can inform the development of more targeted, position-specific training protocols to enhance the overall performance and endurance of field hockey teams.

References

Bangsbo J (1994) The physiology of soccer–with special reference to intense intermittent exercise. Acta Physiol Scand Suppl 619:1–155

Bangsbo J, Lindquist F (1992) Comparison of various exercise tests with endurance performance during soccer in professional players. Int J Sports Med 13(2):125–132

Bangsbo J, Iaia FM, Krustrup P (2008) The Yo-Yo intermittent recovery test. Sports Med 38(1):37–51

Bramell A (2016) Differences in the relationship of heart rate and blood lactate values between running on track versus inclined treadmill: improving laboratory testing to accurately prescribe exercise intensities 1–31
Calo CM, Sanna S, Piras IS, Pavan P, Vona G (2009) Body composition of Italian female hockey players. Biol Sport 26(1):23
Elferink-Gemser MT, Visscher C, Lemmink KA, Mulder T (2007) Multidimensional performance characteristics and standard of performance in talented youth field hockey players: a longitudinal study. J Sports Sci 25(4):481–489
Elferink-Gemser MT, Visscher C, van Duijn MA, Lemmink KA (2006) Development of the interval endurance capacity in elite and sub-elite youth field hockey players. Br J Sports Med 40(4):340–345
Ferry A, Duvallet A, Rieu M (1988) The effect of experimental protocol on the relationship between blood lactate and workload. J Sports Med Phys Fitness 28:341–347
Fox EL, Bowers RW, Foss ML (1988) The physiological basis of physical education and athletics, 4th edn. WB Saunders Company, Philadelphia
Ghosh AK, Goswami A, Mazumdar P, Mathur DN (1991) Heart rate and blood lactate response in field hockey players. Indian J Med Res 94:351–356
Gladden LB (2004) Lactate metabolism: a new paradigm for the third millennium. J Physiol 558(1):5–30
Guerra RS, Amaral TF, Marques E, Mota J, Restivo MT (2010) Accuracy of Siri and Brozek equations in the percent body fat estimation in older adults. J Nutr Health Aging 14(9):744–748
Helgerud J, Engen LC, Wisloff U, Hoff JAN (2001) Aerobic endurance training improves soccer performance. Med Sci Sports Exerc 33(11):1925–1931
Hermansen L, Osnes JB (1972) Blood and muscle pH after maximal exercise in man. J Appl Physiol 32(3):304–308
Ismail KAYA, Kalkavan A, Senturk A, Harmanci H, Karavelioglu M, Yuksel O, Savasli M (2013) Analysis of blood lactate and heart rate of indoor sports athletes as a response to various loads. Türk Spor ve Egzersiz Dergisi 15(3):86–89
Janssen P (2001) Training at anaerobic threshold. Lactate threshold training, 2nd edn, vol 4. Human Kinetics, United States, pp 25–41
Lopes RF, Osiecki R, Rama LMPL (2012) Heart rate and blood lactate concentration response after each segment of the Olympic Triathlon event. Revista Brasileira de Medicina do Esporte 18:158–160
Mann T, Lamberts RP, Lambert MI (2013) Methods of prescribing relative exercise intensity: physiological and practical considerations. Sports Med (Auckland, N.Z.) 43(7):613–25
Meyer T, Gabriel HH, Kindermann W (1999) Is determination of exercise intensities as percentages of VO_2max or HRmax adequate? Med Sci Sports Exerc 31(9):1342–1345
Montgomery DL (2006) Physiological profile of professional hockey players-a longitudinal comparison. Appl Physiol Nutr Metab 31(3):181–185
Morgan W (1973) Psychological factors influencing perceived exertion. Med Sci Sports 5(2):97–103
Muth ND (2009) What are the guidelines for percentage of body fat loss? American Council on Exercise (ACE). Ask the Expert Blog
Norton K, Olds T, Olive S, Craig N (1996) Anthropometry and sports performance. In: [3] Norton K, Olds T (eds) Anthropometrica. University of New South Wales Press, Sydney, Australia
Ohkuwa T, Tsukamoto K, Yamai K, Itoh H, Yamazaki Y, Tsuda T (2009) The relationship between exercise intensity and lactate concentration on the skin surface. Int J Biomed Sci IJBS 5(1):23
Pasco JA, Holloway KL, Dobbins AG, Kotowicz MA, Williams LJ, Brennan SL (2014) Body mass index and measures of body fat for defining obesity and underweight: a cross-sectional, population-based study. BMC Obesity 1(1):1–7
Reilly T, Borrie A (1992) Physiology applied to field hockey. Sports Med 14(1):10–26
Sharma HB, Kailashiya J (2017) Anthropometric and physiological basis of endurance capacity in Young Indian Field Hockey Players. Indian J Physiol Pharmacol 61(2):114–121

Tamer K (1991) Fiziksel performansın ölçülmesi ve değerlendirilmesi. Egzersiz Fizyolojisi Laboratuvar Rehberi) Gökçe offset matbaacılık Ankara S 32

Ucan Y (2015) Effect of national-level field hockey on physical fitness and body composition parameters in Turkish females. Contemp Sports Issues Sports Nutr Sports Stud Sports Psychol Women Sports 5(6):12–16

Ural ZF (1972) Koruyucu Hekimlik Hijyen ve Sanitasyon, 5th edn. Ankara University Publishing, Ankara

Veerappa M (2018) Indian women's hockey players acing the yo-yo test. Retrieved from https://timesofindia.indiatimes.com/sports/hockey/top-stories/indian-womens-hockey-players-acing-the-yo-yo-test/articleshow/62482398.cms. Accessed 12 Jan 2022

Vergès S, Flore P, Favre-Juvin A (2003) Blood lactate concentration/heart rate relationship: laboratory running test versus field roller skiing test. Int J Sports Med 24(06):446–451

Wagner PD (1996) A theoretical analysis of factors determining VO2max at sea level and altitude. Respir Physiol 106(3):329–343

Wasserman K, Beaver WL, Whipp BJ (1986) Mechanisms and patterns of blood lactate increase during exercise in man. Med Sci Sports Exerc 18:344–352

Wood C (2021) Hockey player body size. Topend Sports Website. Retrieve from https://www.topendsports.com/sport/hockey/anthropometry-height-weight.htm. Accessed 23 Sept 2021

Chapter 10
Body Fat and Bone Mineral Density of Endurance Sports Persons and Resistance Training Persons

Naseer Ud Din Waza and Nishan Singh Deol

1 Introduction

The physical activity is thought to be a preventive factor that can reduce the risk of osteoporosis however, not all forms of exercise have the same beneficial effects on bone health. Swimming is a non-weight-bearing sport that could have a negative or no impact Bellver et al. (2019). Numerous studies have shown the benefits of weight-bearing workouts for long-term bone health. However, compared to other athletes, many endurance athletes are at a disadvantage when it comes to bone health due to specific changes in bone-loading characteristics and energy expenditure during involvement in endurance sports. As compared to athletes who compete in weight-bearing sports, people who engage in low-impact, endurance sports like running and non-weight-bearing sports like cycling and swimming frequently have low bone mineral density (BMD). Low BMD increases the risk of stress and fragility fractures, both during athletic activity and later in life (Scofield and Hecht 2012).

Body composition entails analysing human body on the basis of fractionation of total body mass. In sports, the body composition assessment plays a vital role since it can ascertain the athletic potential and success in a particular sport combined with functional, physical and psychosocial factors (Gil 2007; Ramos-Sepulveda 2012). In sports, body fat must be observed, as suitable levels of fat ensure effective movement of players during training and games. Lean mass, particularly muscle mass, must also be observed, as unsuitable training loads can cause unwanted fluctuations in physique, which could affect speed, power, strength and risk of injury (Collins and Rollo 2014; Sutton et al. 2009).

Sports activity type can have significant changes on body composition. Each particular sports discipline requires a specific type of training and physical activity that clearly affects athlete's body composition; so, applying a rigid notion of optimal

N. U. D. Waza (✉) · N. S. Deol
Department of Physical Education, Punjabi University, Patiala, India
e-mail: naseer_rs19@pbi.ac.in

A. D. Singh et al. (eds.), *Sports Ergonomics*, Design Science and Innovation,
https://doi.org/10.1007/978-981-97-7804-1_11

body composition to every sport is not possible (Thomas et al. 2016). Strategies to achieve the best possible physique for each sports discipline form part of a comprehensive method to explore the performance of the athletes. To achieve this purpose, recent studies conducted on athletes emphasized the need of reference data for evaluating the body composition of athletes of various sports, and with or without physical impairments (Micheli et al. 2014; Campa and Toselli 2018). In the absence of suitable references for each sport group, researchers continue to use reference standards from the overall healthy population (Pollastri et al. 2016).

2 Methodology and Procedure

2.1 Selection of Subjects

To achieve the purpose of the study, a total of 30 university level male participants, with endurance sports group ($n = 15$) and resistance training group ($n = 15$), were selected as subjects. In endurance sports, athletes participating in long distance running were selected and categorized as endurance group. For resistance training, weight lifters and body building persons were selected and categorized as resistance training group.

2.2 Inclusive and Exclusive Criteria

The participants included in the study were healthy and free from any musculoskeletal problems. Any subject having any health problem and medical condition was excluded from the study.

3 Selection of Variables

The following variables were selected to accomplish the objectives of the study.

a. Fat mass (kg)
b. Bone mineral density (Stiffness Index)

4 Instruments Used

1. Body composition analyser GS6.5B Body Building Weight Test System (Version 1.0)

2. Bone mineral density analyser Achilles EXPII (Part 1)

4.1 *Criteria Measure*

a. **Body Composition Analyser**

The body composition analyser was used to measure body fat content of the defined population.

b. **Bone Mineral Density Analyser**

The bone mineral density analyser was used to examine bone mineral density of defined population.

5 Analysis and Interpretation of the Data

Table 1 gives the mean and standard deviation (MEAN ± SD) with regards to the bone mineral density of endurance athletes which was 109.60 ± 12.51 and 115.73 ± 7.06 for resistance training group. The t-value with regards to the bone mineral density was 1.6537 and *p*-value was 0.1093 and was not found to be statistically significant at 0.05 level of significance $p > 0.05$. However, the mean value of resistance training group for bone mineral density was higher than that of endurance athletes group (Fig. 1).

Table 2 gives the mean and standard deviation (MEAN ± SD) with regards to the variable body fat of endurance athletes was 10.213 ± 1.996 and 15.52 ± 6.159 for body builders. The *t*-value with regards to the body fat was 3.1746 and *p*-value was 0.0036 and was found to be statistically significant at 0.05 level of significance $p < 0.05$ (Fig. 2).

Table 1 Descriptive statistics of bone mineral density of endurance athletes and resistance training group

Groups	N	MEAN	SD	SEM	*T* value	*P* value
Endurance athletes	15	109.60	12.51	3.23	1.6537	0.1093
Resistance training group	15	115.73	7.06	1.82		

Table 2 Descriptive statistics of body fat of endurance athletes and resistance training group

Groups	N	MEAN	SD	SEM	*T* value	*P* value
Endurance athletes	15	10.213	1.996	0.515	3.1746	0.0036
Resistance training group	15	15.52	6.159	1.590		

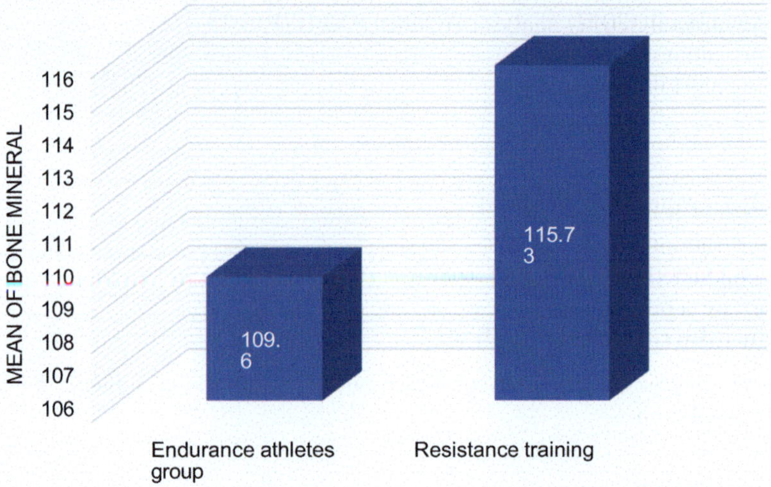

Fig. 1 Showing the mean of bone mineral density of endurance athletes and resistance training group

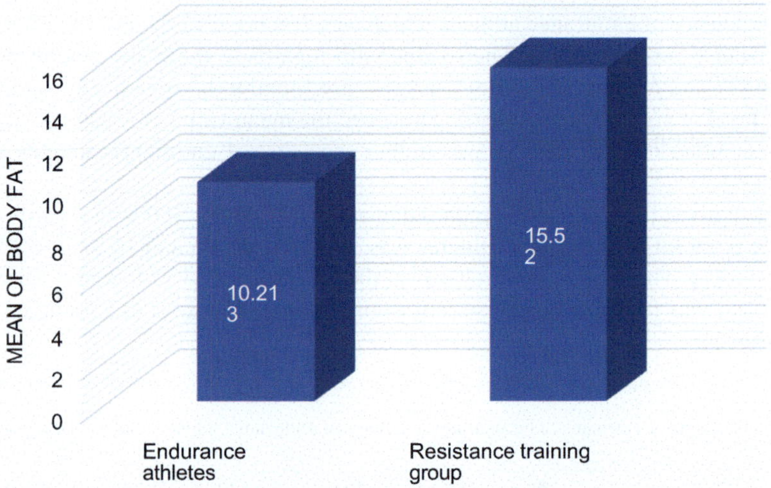

Fig. 2 Showing the mean of body fat for endurance athletes and body builders

6 Results

The study found that the variable bone mineral density does not vary significantly between endurance sports persons and resistance training persons. However, the mean value of resistance training group for bone mineral density was higher than that of endurance athletes group. The results are supported by the study conducted by

Scofield and Hecht (2012). During their study, they found that athletes who participate in various endurance sports activities such as running and athletes participating in swimming and biking have lower bone mineral density than those participating in ball and power sports. Another study conducted by Sagayama et al. (2020) concluded that weight-classified athletes have significantly higher bone mineral density compared to endurance athletes.

This study found that there exists a significant difference in body fat among resistance training group and endurance sports persons. The results are supported by a study conducted by Willis et al. (2012) who found that while resistance training is necessary for a programme to increase lean mass in middle-aged, overweight/obese people, aerobic training is the best kind of exercise for lowering fat mass and body mass.

7 Conclusion

This study found that the variable bone mineral density does not vary significantly between endurance sports persons and resistance training persons. However, the mean value of resistance training group for bone mineral density was higher than that of endurance athletes group. Athletes with low bone mineral density should modify their exercise programmes to include weight training and jumping sports to improve the bone mineral density. To maintain a healthy body mass and body fat a combination, training programme consisting of aerobic and resistance training should be followed.

References

Bellver M, Del Rio L, Jovell E, Drobnic F, Trilla A (2019) Bone mineral density and bone mineral content among female elite athletes. Bone 127:393–400

Campa F, Toselli S (2018) Bioimpedance vector analysis of elite, sub elite, and low-level male volleyball players. Int J Sports Physiol Perform 13:1250–1253

Collins J, Rollo I (2014) Practical considerations in elite football. Sports Sci Exchange 27(133):1–7

Gil SM (2007) Physiological and anthropometric characteristics of young soccer players according to their playing position: Relevance for the selection process. J Strength Cond Res 21:438–445

Micheli ML, Pagani L, Marella M, Gulisano M, Piccoli A, Angelini F, Gatterer H (2014) Bioimpedance and impedance vector patterns as predictors of league level in male soccer players. Int J Sports Physiol Perform 9(3) 532-–539

Pollastri L, Lanfranconi F, Tredici G, Schenk K, Burtscher M, Gatterer H (2016) Body fluid status and physical demand during the Giro d'Italia. Res Sports Med 24(1):30–38

Ramos-Sepulveda JA (2012) Anthropometric and fitness indicators for the selection of young Vallecaucan soccer players. Ludica Pedagog. 2:120–129

Sagayama H, Kondo E, Tanabe Y, Ohnishi T, Yamada Y, Takahashi H (2020) Bone mineral density in male weight-classified athletes is higher than that in male endurance- athletes and non-athletes. Clinical Nutrition ESPEN 36:106–110

Scofield KL, Hecht S (2012) Bone health in endurance athletes: runners, cyclists, and swimmers. Curr Sports Med Rep 11(6):328–334

Sutton L, Scott M, Wallace J, Reilly T (2009) Body composition of English Premier League soccer players: influence of playing position, international status, and ethnicity. J Sports Sci 27(10):1019–1026

Thomas DT, Erdman KA, Burke LM (2016) American College of Sports Medicine joint position statement. Nutrition and athletic performance. Med Sci Sports Exerc 48:543–568

Willis LH, Slentz CA, Bateman LA, Shields AT, Piner LW, Bales CW, Kraus WE (2012) Effects of aerobic and/or resistance training on body mass and fat mass in overweight or obese adults. J Appl Physiol

Chapter 11
Antioxidant and Oxidative Stress in Sports Personnel

Sweta Goyal and Tarvinder Jeet Kaur

1 Introduction

1.1 Physiology of Free Radicle

Free radicles are highly reactive species because they contain unpaired electrons (Lobo et al. 2010). The most important radicals are reactive oxygen species (ROS) and reactive nitrogen species (RNS). There are four ROS: singlet oxygen ($^1O_2$), superoxide ($O^{\bullet -}$), hydroxyl radicle ($^{\bullet}OH$), and hydrogen peroxide (H_2O_2). Peroxynitrite ($ONOO^-$) and nitric oxide ($^{\bullet}NO$) are included as RNS. When superoxide disintegrates into oxygen and hydrogen peroxide in the presence of iron, highly reactive hydroxyl radicals are generated whereas nitric oxide is generated by nitric oxide synthase (NOS) Fig. 1. The production of free radicals is a natural process and intracellularly they are the side-product of the metabolism mitochondrial respiratory chain (Sanderson et al. 2013) while extracellularly, they are generated as a part of the immune response by NADPH oxidases (NOX), myeloperoxidases (MPO) and xanthine oxidases (XO) (Kahles and Brandes 2012; Vergeade et al. 2012). The external triggers for the production of free radicals are pollution, UV rays, drugs, cosmetics, xenobiotics, etc.

Although regulated production of free radicals is a result of normal physiological process that aids in the metabolic process and the immune response. There are many antioxidant defense systems in the cell to counter the production of ROS or to remove them such as glutathione peroxidase (GPx), glutathione, catalase (CAT), superoxide dismutase (SOD), retinol, and vitamin E (Fig. 2).

S. Goyal (✉) · T. J. Kaur
Department of Home Science, Kurukshetra University, Kurukshetra, Haryana, India
e-mail: sweta.homesc@kuk.ac.in

A. D. Singh et al. (eds.), *Sports Ergonomics*, Design Science and Innovation,
https://doi.org/10.1007/978-981-97-7804-1_12

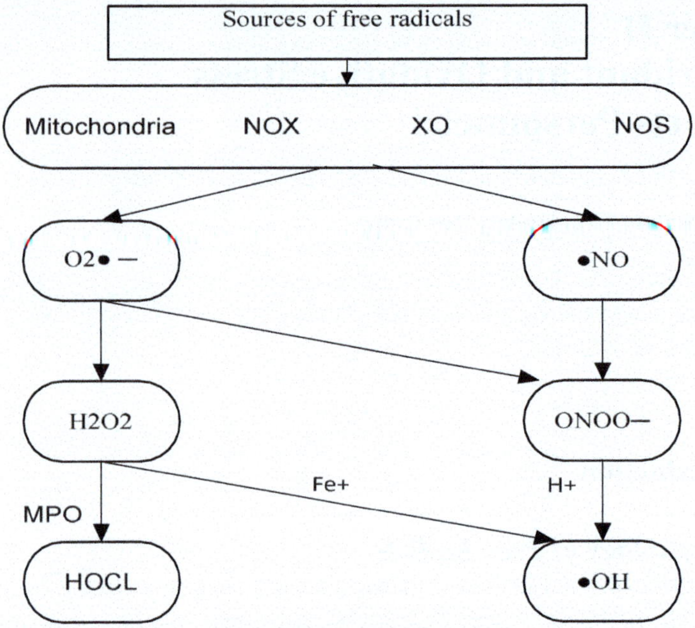

Fig. 1 Mechanism involved in the production of free radical

Fig. 2 Enzymatic and non-enzymatic antioxidants and their targeted free radicals

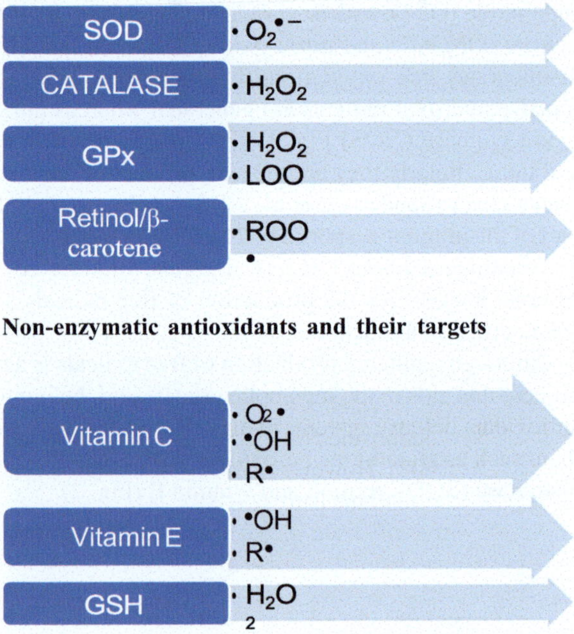

1.2 Oxidative Stress

Negative balance between the generation of oxidants and the in-built mechanism of neutralizing them by antioxidants in favor of oxidants is called oxidative stress. Oxidants can modify lipids, proteins, carbohydrates, and nucleic acid. The formation of free radicals is primarily initiated by the utilization of molecular oxygen, which is due to its structure, is in fact a free radicle itself (Shadab et al. 2014). Oxidative stress leads to oxidative damage, stress-sensitive signaling pathway, and the development of different diseases such as metabolic syndrome, neurological disorders, cardiovascular diseases, cancer, and insulin resistance (Otani 2011; Valko et al. 2007). It is also demonstrated that ROS causes premature fatigue during exercise (Falone et al. 2010).

1.3 Biomarkers that Reflect In Vivo Oxidative Stress

ROS causes oxidative damage to biomacromolecules such as protein, carbohydrates, lipids, and DNA. These oxidative damage products are known as biomarkers. It is very difficult to measure the concentration of ROS directly because of their extremely short life span. Therefore, products generated from oxidative damage are often used to quantify the level of oxidative damage. By the determination of these biomarkers, useful information can be deduced for avoiding certain diseases.

Powers and Jackson mention four categories to determine oxidative stress in their review. The first method is, detection of the quantity of free radicals. As mentioned earlier, it is very difficult to detect them because of their short life span. To use exogenous molecules such as fluorescent probes or spin traps (i.e., N-tert- butyl -α-phenylnitrone and 5,5-dimethyl-1-pyrroline-N-oxide) to produce luminescence or to stabilize these molecules is the only way to measure them. The second method is, detection of antioxidant activity or antioxidant level in the tissue. The antioxidant level may decrease or increase according to the degree of exposure to oxidative stress present in body. Although antioxidants are a very useful biomarker, but they are affected by nutritional status or diet. The third method is, measurement of the oxidation products, namely protein carbonyl (PC), the marker of protein oxidation, $F2$-isoprostane, and malondialdehyde (A), the marker of lipid peroxidation, and 8-oxo-20 deoxyguanosine (8-OH-dG), which is a marker of DNA oxidation (Powers and Jackson 2008). However, these products exist only in trace amounts, so they are often problematic to measure. The fourth method is, measurement of redox balance. The most commonly used redox marker is reduced glutathione/oxidized glutathione ratio, cysteine redox state, and thiol/disulfide. Malondialdehyde (MDA) (by-product of lipid peroxidation), advanced oxidation protein product (AOPP), and ImAnOx are the markers of total antioxidant activity. All these biomarkers are measured by ELISA assay kit, spectroscopic analysis AOPP kit, and photometric test system ImAnOx kit, respectively.

2 Material and Methods

Facts and particulars related to impact of antioxidant supplementation on sports personnel were searched through internet/digital libraries in all English-language publications. All published articles from 2000 to 2022 were reviewed by searching the database of PubMed, Google Scholar, Scopus, Web Of science, and ScienceDirect. Articles were searched using English keywords such as 'biomarkers of oxidative stress', 'oxidative stress on sports personnel', and 'impact of supplementation on oxidative stress biomarkers'.

3 Result and Discussion

3.1 Exercise-Induced Oxidative Stress in Sports Personnel

During physical exercise, oxygen demand increases for the activation of skeletal muscles and triggers the generation of ROS which attack biological macromolecules especially fatty acid, amino acids, and DNA, (Lambertucci et al. 2007). During the exercise, metabolic rate and temperature increases as well as utilization of O_2 by muscle fibers increases. All these factors combined with the low pH level of cellular muscle which accelerate the production of ROS. Angelo and Rosa concluded that the production of free radicals increases with an increase in oxygen consumption and oxygen phosphorylation (D'Angelo and Rosa 2020). Major sources of ROS generation in active muscles during exercise are mitochondria, xanthine oxidase, nicotinamide adenine dinucleotide phosphate (NADPH) oxidase, and some immune cells such as macrophages, monocytes, eosinophils, and neutrophils (Powers et al. 2011; Kinnunen et al. 2005).

Another reason of ROS generation is increased levels of catecholamines and metmyoglobin from damaged muscle and their interaction with peroxides during exercise (Powers and Jackson 2008; Cooper et al. 2002). Another study reviewed that the contraction of muscles is also the active site of the production of ROS (Power et al. 2020). Moreover, for many years, only skeletal muscles had been interpreted as a source of ROS but in the recent years more sites like blood (hemoglobin and myoglobin) have been recognized (Cooper et al. 2002). A study showed that the increase in the level of serum Malondialdehyde (MDA) after an intense bout exercise increase in oxidative stress and it leads to decrease in the performance of sports personnel by reduction of contraction of the muscles, thus producing fatigue (Bouzid et al. 2014). Lu et al. reviewed that oxidative stress is positively related to the duration and intensity of high-intensity exercise (HIE). They further added that the degree of oxidative stress is also affected by individual's exercise habits, fitness level, age, and gender. Physically active people have greater antioxidant capacities however, on the contrary, they observed older people have been observed less susceptible to oxidative damage than younger people (Bouzid et al. 2014). Males are more prone to develop oxidative stress compared to females (Lu et al. 2021) while Wiecek et al. noted that gender played no role in the development of oxidative stress (Wiecek et al. 2018). Djordjevic et al. compared before-and-after exercise antioxidant level in young handball players and concluded that at rest players had high level of SOD, GSH, NO, and CAT activity and lower level of lipid peroxidation after exercise there is fall in NO and CAT activity as well as $O2^{\bullet}$ (Djordjevic et al. 2012). A study, done by Jammes et al., on athletes doing cycling found that maximal thiobarbituric acid reactive substance (TBARS) was positively correlated to oxidative stress with VO2max. (maximal oxygen uptake) (Jammes et al. 2004). Ammar et al. studied that aerobic, anaerobic, and combined exercise have the potential to increase oxidative stress and antioxidant activities depending upon the duration and intensity of physical exercise (Ammar et al. 2020). Groussard et al. concluded that short duration supramaximal anaerobic exercise could induce oxidative stress (Groussard et al. 2003). While according to Bogdanis et al. short duration high-intensity exercise decreases oxidative stress and regulates antioxidant activity after only nine sessions for 20 min (Bogdanis et al. 2013) (Table 1)

3.2 *Exercise and Endogenous Antioxidant Defense System*

Some studies showed that free ROS generated during regular and moderate exercise function as hormesis and activate endogenous antioxidant defense system by activating some components such as superoxide dismutase and glutathione peroxidase (Gomez-Cabrera et al. 2008; Ji 2002). So acute and intense exercise produces ROS, RNS and oxidative stress that have a negative impact on skeletal muscles (Belviranlı and Gökbel 2006) but, non-exhaustive physical exercise induces an endogenous antioxidant system and defends the body against oxidative damage (Reid 2016; Kruk et al. 2019). Falone et al. studied habitual aerobic male runners and concluded that long-term regular and moderate aerobic exercise can increase the antioxidant defense system (Falone et al. 2010). This increase may be a consequence of an improvement in the plasma activity of CAT, GPx, SOD, and GR (Berzosa et al. 2011). Miyazaki et al. studied that high-intensity endurance training could elevate the antioxidant enzyme level in erythrocytes which decreases $O^{\bullet}$ production (Miyazaki et al. 2001), and could protect from chronic oxidative stress-associated diseases (Parker et al. 2014). A study, done on 10 men and 10 women performed 20-s bicycle sprints, showed the changes in antioxidants level, antioxidant status (TOS), and antioxidant activity (TAC). Samples were taken between 3 min and 24 h. They found that TOS, TAC, vitamin A, and vitamin E concentration were the highest at 3 min, while Vit C and uric acid were the highest at 1 h. GSH/GSSH ratio was significantly lower at 1 h and 24 h than the before exercise after anaerobic exercise (Wiecek et al. 2018).

Table 1 Studies related to oxidative stress biomarkers after exercise

Participants	Exercise	Duration	Result	Study
Athletes $n = 11$ (males 8 female 3)	50 km ultramarathon		*F*2-isoprostane↑ *A*-tocopherol↑ Ascorbic acid↑	Mastaloudis et al. (2001)
Male triathletes $n = 18$	Swimming 3.8 km + then a 180 km bike ride, + finished with a 42.2 km marathon run		TBARS↑ PC↑ SOD↑ CAT↑ LHP↑	Pinho et al. (2010)
Male athletes $n = 42$	3.8 km swimming + 180 km cycling + 42.2 km running	2 days pre, 1st, 5th, 19th post-ex	MDA↑ *A* and γ tocopherol↑, β-carotene↑	Wagner et al. (2010)
Male athletes $n = 21$	Two successive editions of ultramarathon for 330 km, (altitude difference of 24,000 m distance was divided into seven stages with 35 relax stations)	150 h	8-OH-dG↑ 8-iso-PGF2α↑ Antioxidant activity↓	Mrakic-Sposta et al. (2015)
Runner $n = 24$	Double ring 50 km-long and 100 km-long competitive ultramarathon race	4 h 3 minand 8 h 42 min	PC↑ ROS↑ TAC↑ TBARS↑ 8-OH-dG↑	Vezzoli et al. (2016)
Runner $n = 12$	103 km mountain ultra-marathons	72 hr	GSH↓ PC ↔ TAC ↔ CAT ↔ TBARS ↔	Spanidis et al. (2017)
Participants $n = 20$	Marathon	4th and 7th day after ex	8-OH-dG↑ 8-OxoGuo↑	Larsen et al. (2020)

3.3 Antioxidant and Exercise-Induced Oxidative System

Antioxidants break the chain reaction of free radicals by giving electrons and preventing oxidative stress-related damage. According to their activity, antioxidants are of two types—enzymatic and non-enzymatic (Banafsheh and Sirous 2016). Enzymatic antioxidant functions by converting the oxidized metabolic product into hydrogen peroxide and then into the water using cofactors such as iron, zinc, copper, and manganese (Deponte 2013). Several enzymes inhibit free radicals' formation. Primary enzyme act directly to scavenge ROS, whereas secondary enzyme indirectly supports endogenous antioxidant system. For example, glucose-6-phosphate dehydrogenase regenerates NADPH (Banafsheh and Sirous 2016).

The non-enzymatic antioxidants are low molecular weight substances, non-catalytic, and need to supply from the diet. These are of two types, endogenous which are synthesized by eukaryotic cell and exogenous which are necessary to take through the diet. Endogenous antioxidants play a significant role in maintaining optimal cellular functions. The most efficient enzymatic oxidants are glutathione peroxidase (GPx), catalase (CAT), and superoxide dismutase (SOD) and non-enzymatic antioxidants include vitamin E, A, and C, thiol antioxidant (glutathione, thioredoxin, and lipoic acid), ubiquinone, α-lipoic acid, melatonin, uric acid, carotenoids, curcumin, flavonoids, allyl sulfide, bilirubin, polyamines, coenzyme Q (10), etc. (Díaz-Castro et al. 2012; Mirończuk-Chodakowska et al. 2018; Nimse and Palb 2015). When oxidative stress increases and the endogenous antioxidant system is not sufficient, dietary antioxidants are required to maintain optimal cellular function. An antioxidant should possess these qualities like easily absorbed, neutralize free radicals, and chelate redox metals.

3.4 Effect of Antioxidant Supplementation on Oxidative Stress (Exercise-Induced)

When the endogenous antioxidant system is not quite enough to regulate free radicals, then an additional oral antioxidant supplement may work as a non-invasive tool for preventing oxidative stress after exercise. The antioxidant supplement may be warranted when a sports person exposed to severe oxidative stress during intensive training or fail to meet the dietary antioxidant requirement. A study conducted by Ponce-Gonzalez et al. studied three teams of soccer players and found that the winning teams have higher antioxidant activity level. They concluded that it is necessary to check antioxidant activity level to increase success rate by providing diet, training, and ergogenic aid accordingly (Ponce-Gonzalez 2021). A study done on mice that were fed α-tocopherol (200 mg/kg/d) for 13 weeks, showed that supplementation decreased the rate of mitochondrial superoxide ion radical ($O2^{\bullet -}$) generation (Lass and Sohal 2000) (Table 2)

An excess of antioxidants supplement in the body may have detrimental effects on health and performance. Thus, an intake of antioxidants through a balanced diet remains the best way to maintain an optimal antioxidant level. Antioxidants are found in natural sources such as fruits, vegetables, and meat (Grozea 2012). They can be obtained through different parts of plants like nuts, seeds, leaves, bark, and roots (Akbarirad et al. 2016). Some studies have shown the positive effects of natural sources of antioxidants in reducing oxidative stress (Table 3). Fagan

Table 2 Effect of the antioxidant supplement on oxidative stress (Human trial)

Participant	Exercise	Supplementation on	Duration	Result	Stud y
Endurance athletes man $n = 8$	Cycling for 45 min at 92 % VO_{2max} Peak	N-acetylcysteine 125mg/kg/h for 15 min and 25 mg/kg/h for 20 min	45 min (pre-ex and throughout exercise)	↑Muscles glutathione and total glutathione	Medved et al. (2004)
Healthy man $n = 18$	70 eccentric elbow extensions	Vit C 3g/d	18 days (2 weeks pre-ex and 4 post-ex	↓glutathione oxidation	Bryer and Goldfarb (2006)
Male young soccer player $n = 10$	I(30d) aerobic power II(30d) strength capacity II (30d) improving speed and anaerobic power	250 mg ascorbic acid and 200 mg of α-tocopherol four times a day	90 days	↑ LDH, TBARS	Zoppi et al. (2006)
Male $n = 14$	Static bicycle on 3days/week	1g/d Vit C	8 weeks	↓SOD, GPx	Gomez-Cabrera et al. (2008)
Male volunteer $n =$ 21	Elbow flexion and extension at an intensity of 80% VO_{2max} for maximum repetition	800 IU per day D-α- tocopherol	21 (14 days pre- ex and 7 days post-ex)	↑ LDH, TBARS, protein carbonyl,	Silva et al. (2010)
Trained (Moderately) males $n = 22$	Downhill (40 min) running at 65–70% VO_{2max}	Vit C 1000mg	2 weeks	↑antioxidant capacity	He et al. (2015)
Resistance trained male $n = 9$	Two RE bouts pre and post intervention using dynamometer	250 mg/12 hr vit C	28 days	↓oxidative stress, MDA.	Evans et al. (2017)
Recreation al active athletes $n = 20$	4x30-s bouts of cycling	Two 500mg perfload® polyphenol-based food supplement.	60 min	↑SOD, GPx, CAT, LDH	Cases et al. (2017)

(continued)

Table 2 (continued)

Participant	Exercise	Supplementation on	Duration	Result	Stud y
Males (different glutathione level) *n* = 100	Aerobic exercise till exhaustion	600 mg twice daily	30 days	↑glutathione concentration	Paschalis et al. (2018)
Healthy man (physically active) *n* = 9	Treadmilltest (till exhaustion) under anaerobic and aerobic condition	Vit E 250 mg	One pre-ex hr	↓lactate dehydrogenase	Santos et al. (2016)
Football players *n* = 21	Plyometric jumping and strength	Vitamin C 500 mg/d and E 400 UI/d	15 d (7 days pre- ex 8 days post-ex)	↓oxidative stress, MDA, total lipid peroxidation, glutathione/oxidized	de Oliveira et al. (2019)
	resistancetests to exhaustion			glutathione	
Male soccer players (*n* = 13)	Running exercise until exhaustion	6 melatonin mg	48 hrs	↓Oxidation protein product, biomarkers of muscles, liver, and renal damage ↑antioxidant enzyme activity	Farjallah et al. (2021)
Male soccer players (n = 20)	6 days exhaustive training schedule	5 melatonin mg	6 days	↓Oxidation protein product, biomarkers of muscles, liver, and renal damage ↑antioxidant enzyme activity	Farjallah et al. (2022)

et al. compared the effectiveness of synthetic source and natural source of α- tocopherol (4000 IU/d) and found α-tocopherol from natural source was more efficient in reducing lipid peroxidation (Fagan et al. 2020). In a 20 weeks study Roberts II et al. also found that RRR-d-α-tocopherol acetate (3200 IU/d), from a natural source, reduced F2-isoprostanes which is a biomarker of oxidative stress (Roberts II et al. 2007). According to a study, the intake of red wine (240 ml/d) reduced 8-hydroxy-2'-deoxyguanosine (8-OH-dG) and plasma nitro tyrosine, a biomarker of oxidative stress (Urquiaga et al. 2010) (Table 4).

4 Conclusion

ROS generation is a natural process and a result of normal cell activity and participation in cellular signaling and a by-product of oxygen metabolism. The endogenous antioxidant defense system works as a big system to maintain redox balance in the body. Moderate exercise can exceed the capacity of the endogenous antioxidant defense system while intense exercise increases the oxidative stress in sports personnel and the endogenous defense system is unable to neutralize this excess of ROS, then it is compulsory to take antioxidant orally. Despite remarkable evidence of the positive effect of vitamins supplements, vitamins-rich diet has been proved to be improved oxidative stress along with maintaining antioxidant defense system equilibrium. Hence, it is necessary to take antioxidants through diet to maintain the antioxidant level in the body safely.

So, the best recommendation regarding antioxidant supplementation is to have this from a natural source with a balanced diet rich in antioxidants. Regular consumption of whole grains and millet, legumes and beans, fruits and vegetables, nuts and seeds, and green tea and dark chocolate is an effective and safe way to meet the requirement of athletes and sports personnel.

Table 3 Effect of dietary antioxidant on oxidative stress

Participant	Exercise	Supplementation on	Duration *n*	Result	Stud *y*
Moderate training cyclist $n = 30$ Placebo $n = 30$	Submaximal 90 min aerobic exercise on bicycle	Beverage (81 ml g/l black grapes, 36 g/l red current, 93 g/l raspberry) 2.3 g polyphenol/trial	After 45 min of exercise	↓lipid peroxidation, carbonyl groups (CO)	Morillas-Ruiz et al. (2006)
Physical education students $n = 14$	Swimming test six repeats of 50 m	3 capsules of 390 mg/d red grape skin extract	6 weeks	↑GSH, uric acid, TAS	Sadowska-Krępa et al. (2008)
Recreational marathon runner $n = 20$	Marathon race	Tart cherry juice	5 days	↑TAC ↓ TBARS level	Howatson et al. (2010)
Trained male rowers $n = 22$	Repeated cycle sprint test on rowing till 90% VO_{2max}	3 times/d penance vid 2000® made with black wine grapes extract	6 weeks	↑TAC, ↓GPx, lipid peroxidation product	Skarpańska-Stejnborn et al. (2010)
Physically active man $n = 20$	Cycling for 90 min up to exhaustion at 90% VO_{2max}	Dark chocolate twice daily one pre-ex	2 weeks	↓blood *F*2-isoprestance, oxidized LDL, FFA	Allgrove et al. (2011)
Male sprinters $n = 16$	Two repeat cycle tests 4 consecutive 15-s bout test 4 weeks of strength training	980 mg/d polyphenol through green tea extract	4 weeks	↑ total polyphenol, TAC ↓ (SQ), and MDA	Jówko et al. (2011)
Recreationally active male $n = 20$	30 min treadmill exercise 50% (Wmax)	500 ml/d POMj	1 week pre-ex	↓ TBARS level	Tsang et al. (2011)
Healthy men $n = 14$	Cycled for 2.5 h at 60% maximal oxygen uptake	100 mg dark chocolate	2 h pre- ex	↑TAC ↓oxidative stress ↓ *F*2-isoprestance	Davison et al. (2012)

(continued)

Table 3 (continued)

Participant	Exercise	Supplementation on	Duration *n*	Result	Stud *y*
Endurance athlete males $n = 28$	Treadmill, runs (max HR 70%)	240 ml/d POMj	15 days pre-ex	↑TAC level, SOD, GPx ↓MDA level and matrix metalloproteinases (MMP)	Mazani et al. (2014)
Recreationally active male $n = 45$	Ten sets downhill running and 40 bilateral eccentric elbow flexion	1*or 2* 500 ml/d POMj	8 d (4 pre-ex and 4 post- ex)	Strength recovery (both leg and arm)	Machin et al. (2014)
Male healthy volunteer $n = 14$	Endurance and resistance exercise sessions	POMj	15 days	↑GSH, TAC, GSH ↓MDA, CARB, CAT	Matthaiou et al. (2014)
Male endurance athlete $n = 28$	Severe based endurance activity	240 ml/d of POMj	2 weeks pre-ex	↑SOD, GPx, TAC ↓MDA	Naghizadeh-Baghi et al. (2015)
Male sprinters $n = 16$	Two repeat cycle tests 4 consecutive 15-s bout test	980 mg/d polyphenol through green tea extract	4 weeks	↓ superoxide dismutase (SOD), and MDA ↑ total polyphenol, TAC	Jówko et al. (2015)
Recreational male runner $n = 28$	Two running test to exhaustion at 80% VO2max	10 ml/kg/d purple grape juice	28 days	↑TAC, vitamin A	Toscano et al. (2015)
Endurance trained runner $n = 27$	Half marathon 21.1 km in 2 h	Capsule containing 480mg tart cherry	10 d (2d post-ex)	↑TAC ↓inflammatory markers	Levers et al. (2016)
Female volleyball player $n = 40$		300 mg twice daily Grape seed extract	8 weeks	↓MDA ↑GSH	Taghizadeh et al. (2016)

(continued)

Table 3 (continued)

Participant	Exercise	Supplementation on	Duration *n*	Result	Stud *y*
Male soccer player I $n = 18$ II $n == 18$ III $n = 18$	HIIT program for two weeks (total 10 sessions)	I 450 mg/d green tea extract (GTE) II 450 mg/d sour tea extract (STE) III 450 mg/d maltodextrin	6 weeks	↓malondialdehyde (MDA) level ↑total antioxidant capacity (TAC) level after having STE	Hadi et al. (2017)
Endurance trained males $n = 22$	1 h/d based- endurance training 3 session/wee k	200 ml/d for Ist group and 100 ml/d for IInd group POMj	21 d pre- ex	↑lactate ↓MDA level and CO	Fuster-Muñoz et al. (2016)
Elite weightlifter = 9	Intense weightlifting training session	3*250 ml/d 8 h interval POMj	pre-ex (1 h) and 2 d post-ex	↓LDH ↓aspartate aminotransferase (ASAT) ↑TAC level ↓MDA level	Ammar et al. (2016, 2017)
Trained amateur male athletes $n = 16$	Submaximal cycling at 60% of peak power output	500 mg/d GTE	15 days	↑↓Oxidative stress,	Machado et al. (2018)
Judo athletes $n = 20$	3 rounds of 7 min combats	400 ml/d grape juice	14 days	↓SOD, DNA damage, lipid damage	Júlia et al. (2020)
Recreational male runner $n = 14$	Two running test to exhaustion at 80% VO_{2max}	10 ml/kg/d grape juice	Pre-ex	↑TAC,	de Lima Tavares Toscano et al. (2020)

Table 4 Effect of antioxidant supplementation on oxidative stress (Animal trial)

Participant	Exercise	Supplementation	Duration	Result	Study
Male Wistar rats $n = 24$	5 days/week on a treadmill	vit C 0.24 mg/cm^2 body surface area	8 weeks	↓SOD, GPx	Gomez-Cabrera et al. (2008)
Male Wistar rats $n = 112$	Swimming 10 h	25 mg/kg D-α-tocopherol, or 25 and 50 mg/kg of tocotrienol-rich fraction,	28 days	↑SOD, CAT, GPx ↓lactate, TBARS, protein carbonyl,	Lee et al. (2009)
Young and aged male rats Fischer 344 $n = 28$	80 maximal stretch-shortening contractions per session- Three times weekly	Vit C 2% by weight and 30,000 mg/kg of DL-alpha tocopherol acetate	4.5 Weeks	↓MDA, GSH, H2O2 ↑GPx	Ryan et al. (2010)
Male Wistar rats $n = 32$	15 min of treadmill running/d	2 mg/d of vit E	14 days	↓TBARS	Górnicka et al. (2019)
Horse having Vitamin E deficient diet $n = 18$	6-week training protocol	4000 IU α-Tocopherol natural source	7 weeks (1 week pre-ex and 6 weeks post-ex)	↓TBARS	Fagan et al. (2020)

References

Akbarirad H, Ardabili GA, Kazemeini SM, Khaneghah MA (2016) An overview on some of important sources of natural antioxidants. Int Food Res J 23(3):928–933

Allgrove J, Farrell E, Gleeson M, Williamson G, Cooper K (2011) Regular dark chocolate consumption's reduction of oxidative stress and increase of free-fatty-acid mobilization in response to prolonged cycling. Int J Sport Nutr Exerc Metab 21(2):113–123

Ammar A, Trabelsi K, Boukhris O, Glenn JM, Bott N, Masmoudi L, Hakim A, Chtourou H, Driss T, Hoekelmann A, El Abed K (2020) Effects of aerobic-, anaerobic- and combined-based exercises on plasma oxidative stress biomarkers in healthy untrained young adults. Int J Environ Res Public Health 17:2601

Ammar b A, Turki M, Hammouda O, Chtourou H, Trabelsi K, Bouaziz M, Abdelkarim O, Hoekelmann A, Ayadi F, Souissi N, Bailey SJ, Driss T, Yaich S (2017): Effects of pomegranate juice supplementation on oxidative stress biomarkers following weightlifting exercise. Nutrients 9:819.

Ammar A, Turki M, Hammouda O, Chtourou H, Trabelsi K, Kallel C Bouaziz M, Abdelkarim O, Hoekelmann A, Ayadi F, Driss T, Souissi N, Bailey SJ, Driss T, Yaich S (2016) Pomegranate supplementation accelerates recovery of muscle damage and soreness and inflammatory markers after a weightlifting training session. PLoS ONE 11:e0160305.

Banafsheh AA, Sirous G (2016): Studies on oxidants and antioxidants with a brief glance at their relevance to the immune system. Life Sci 146:163–173

Belviranlı M, Gökbel H (2006) Acute exercise induced oxidative stress and antioxidant changes. Eur J Gen Med 3(3):126–131

Berzosa C, Cebrián I, Fuentes-Broto L, Gómez-Trullén E, Piedrafita E, Martínez-Ballarín E, López-Pingarrón L, Reiter RJ, García JJ (2011) Acute exercise increases plasma total antioxidant status and antioxidant enzyme activities in untrained men. J Biomed Biotechnol 540458

Bogdanis GC, Stavrinou P, Fatouros IG, Philippou A, Chatzinikolaou A, Draganidis D, Ermidis G, Maridaki M (2013) Short-term high-intensity interval exercise training attenuates oxidative stress responses and improves antioxidant status in healthy humans. Food Chem Toxicol 61:171–177

Bouzid MA, Hammouda O, Matran R, Robin S, Fabre C (2014) Changes in oxidative stress markers and biological markers of muscle injury with aging at rest and in response to an exhaustive exercise. PLoS ONE 9:e90420

Bryer SC, Goldfarb AH (2006): Effect of high dose vitamin c supplementation on muscle soreness, damage, function, and oxidative stress to eccentric exercise Int J Sport Nutr. Exerc Metab 16:270–280

Cases J, Romain C, Marín-Pagán C, Chung LH, Rubio-Pérez JM, Laurent C, Gaillet S, Prost-Camus E, Prost M, Alcaraz PE (2017) Supplementation with a polyphenol-rich extract, perfload®, improves physical performance during high-intensity exercise: a randomized, double blind, crossover trial. Nutrients 9(4):421

Cooper CE, Vollaard NB, Choueiri T, Wilson MT (2002) Exercise, free radicals and oxidative stress. Biochem Soc Trans 30(2):280–285

D'Angelo S. Rosa R (2020): Oxidative stress and sport performance. Sport Sci 13(1):18–22

Davison G, Callister R, Williamson G, Cooper KA, Gleeson M (2012) The effect of acute pre-exercise dark chocolate consumption on plasma antioxidant status, oxidative stress and immunoendocrine responses to prolonged exercise. Eur J Nutr 51(1):69–79

de Lima Tavares Toscano L et al (2020) A single dose of purple grape juice improves physical performance and antioxidant activity in runners: a randomized, crossover, double-blind, placebo study. Eur J Nutr 59:2997–3007

de Oliveira DC, Rosa FT, Simões-Ambrósio L, Jordao AA, Deminice R (2019) Antioxidant vitamin supplementation prevents oxidative stress but does not enhance performance in young football athletes. Nutrition 63:29–35

Deponte M (2013) Glutathione catalysis and the reaction mechanism of glutathione- dependent enzymes. Biochimica Et Biophysica Acts 1830:3217–3266

Dıaz-Castro J, Guisado R, Kajarabille N, Garcıa C, Guisado IM, de Teresa C, Ochoa JJ (2012) Coenzyme Q[10] supplementation ameliorates inflammatory signaling and oxidative stress associated with strenuous exercise. Eur J Nutr 51:791–799

Djordjevic DZ, Cubrilo DG, Barudzic NS, Vuletic MS, Zivkovic VI, Nesic M, Radovanovic D, Djuric DM, Jakovljevic VL (2012) Comparison of blood pro/antioxidant levels before and after acute exercise in athletes and non-athletes. Gen Physiol Biophys 31:211–219

Evans LW, Zhang F, Omaye ST (2017) Vitamin C supplementation reduces exercise- induced oxidative stress and increases peak muscular force. Food Nutr Sci 8:812–822

Fagan MM, Harris P, Adams A, Pazdro R, Krotky A, Call J, Duberstein KJ (2020): Form of vitamin E supplementation effects oxidative and inflammatory response in exercising horses. J. Equine Vet. Sci. 91:103103.

Falone S, Mirabilio A, Pennelli A, Cacchio M, Di Baldassarre A, Gallina S, Passerini A, Amicarelli F (2010) Differential impact of acute bout of exercise on redox- and oxidative damage-related profiles between untrained subjects and amateur runners. Physiol Res 59:953–961

Farjallah MA, Graja A, Ben Mahmoud L, Ghattassi K, Boudaya M, Driss T, Jamoussi K, Sahnoun Z, Souissi N, Hammouda O (2021) Effects of melatonin ingestion on physical performance and biochemical responses following exhaustive running exercise in soccer players. Biol Sport 39:473–479

Farjallah MA, Ghattassi K, Kamoun A, Graja A, Ben Mahmoud L, Driss T, Jamoussi K, Sahnoun Z, Souissi N, Zmijewski P, Hammouda O (2022) Melatonin supplementation alleviates cellular damage and physical performance decline induced by an intensive training period in professional soccer players. PLoS ONE 17(9):e0273719

Fuster-Muñoz E, Roche E, Funes L (2016) Effects of pomegranate juice in circulating parameters, cytokines, and oxidative stress markers in endurance-based athletes: a randomized controlled trial. Nutrition 32:539–545

Gomez-Cabrera M-C, Domenech E, Romagnoli M, Arduini A, Borras C, Pallardo FV, Sastre J, Vina J (2008) Oral administration of vitamin C decreases muscle mitochondrial biogenesis and hampers training-induced adaptations in endurance performance. Am J Clin Nutr 87:142–149

Górnicka M, Ciecierska A, Hamulka J, Drywie ME, Frackiewicz J, Górnicki K, Wawrzyniak A (2019) α-Tocopherol protects the heart, muscles, and testes from lipid peroxidation in growing male rats subjected to physical efforts. Oxid Med Cell Longev 8431057

Groussard C, Rannou-Bekono F, Machefer G, Chevanne M, Vincent S, Sergent O, Cillard J, Gratas-Delamarche A (2003) Changes in blood lipid peroxidation markers and antioxidants after a single sprint anaerobic exercise. Eur J Appl Physiol 89:14–20

Grozea MBAI (2012): Antioxidant (Antiradical) compounds. J Bioequival Bioavail 4(6):10000e18

Hadi A, Pourmasoumi M, Kafeshani M, Karimian J, Maracy MR, Entezari MH (2017): The effect of green tea and sour tea (Hibiscus sabdariffa L.) supplementation on oxidative stress and muscle damage in athletes. J Diet Suppl 14(3):346–357

He F, Hockemeyer J, Sedlock D (2015) Does combined antioxidant vitamin supplementation blunt repeated bout effect. Int J Sports Med 36:407–413

Howatson G, McHugh MP, Hill JA, Brouner J, Jewell AP, van Someren KA, Shave RE, Howatson SA (2010) Influence of tart cherry juice on indices of recovery following marathon running. Scand J Med Sci Sports 20:843–852

Jammes Y, Steinberg JG, Brégeon F, Delliaux S (2004) The oxidative stress in response to routine incremental cycling exercise in healthy sedentary subjects. Respir Physiol Neurobiol 144:81–90

Ji LL (2002) Exercise-induced modulation of antioxidant defense. Ann N Y Acad Sci 959:82–92

Ji LL, Gomez-Cabrera MC, Vina J (2006) Exercise and hormesis: activation of cellular antioxidant signaling pathway. Ann N Y Acad Sci 1067:425–435

Jówko E, Sacharuk J, Balasińska B, Ostaszewski P, Charmas M, Charmas R (2011) Green tea extract supplementation gives protection against exercise-induced oxidative damage in healthy men. Nutr Res 31(11):813–821

Jówko E, Długołecka B, Makaruk B, Cieslinski I (2015): The effect of green tea extract supplementation on exercise-induced oxidative stress parameters in male sprinters. Eur J Nutr 54:783–791

Júlia M, Goulart V, Pisamiglio D, Möller G, Dani C, Alves F, Bock P, Schneider C (2020) Effects of grape juice consumption on muscle fatigue and oxidative stress in judo athletes: A randomized clinical trial. An Acad Bras Ciênc 92(4):e20191551

Kahles T, Brandes RP (2012) NADPH Oxidases as therapeutic targets in ischemic stroke. Cell Mol Life Sci 69:2345–2363

Kinnunen S, Atalay M, Hyyppa S, Lehmuskero A, Hanninen O, Oksala N (2005) Effects of prolonged exercise on oxidative stress and antioxidant defense in endurance horse. J Sports Sci Med 4(4):415–421

Kruk J, Aboul-Enein HY, Kladna A, Bowser JE (2019) Oxidative stress in biological systems and its relation with pathophysiological functions: the effect of physical activity on cellular redox homeostasis. Free Radic Res 53:497–521

Lambertucci RH, Levada-Pires AC, Rossoni LV, Curi R, Pithon-Curi TC (2007) Effects of aerobic exercise training on antioxidant enzyme activities and mRNA levels in soleus muscle from young and aged rats. Mech Ageing Dev 128(3):267–275

Larsen EL, Poulsen HE, Michaelsen C, Kjær LK, Lyngbæk M, Andersen, ES, Petersen-Bønding, C, Lemoine C, Gillum M, Jørgensen NR, Ploug T, Vilsbøll T, Knop FK, Karstoft K (2020) Differential time responses in inflammatory and oxidative stress markers after a marathon: an observational study. J Sports Sci 38:1–12

Lass A, Sohal RS (2000) Effect of coenzyme Q10 and a-tocopherol content of mitochndria on the production of superoxide anion radicals. FASEB J 14:87–94
Lee S-P, Mar G-Y, Ng L-T (2009) Effects of tocotrienol-rich fraction on exercise endurance capacity and oxidative stress in forced swimming rats. Eur J Appl Physiol 107:587–595
Levers K, Dalton R, Galvan E, O'Connor A, Goodenough C, Simbo S, Mertens-Talcott SU, Rasmussen C, Greenwood M, Riechman S, Crouse S, Kreider RB (2016) Effects of powdered montmorency tart cherry supplementation on acute endurance exercise performance in aerobically trained individuals. J Int Soc Sports Nutr 22:1–23
Lobo V, Patil A, Phatak A, Chandra N (2010) Free radicals, antioxidants and functional foods: impact on human health. Pharmacogn Rev 4:10
Lu Y, Wiltshire HD, Baker JS, Wang Q (2021) Effects of high intensity exercise on oxidative stress and antioxidant status in untrained humans: a systematic review. Biology 10:1272
Machado ÁS, da Silva W, Souza MA, Carpes FP (2018) Green tea extract preserves neuromuscular activation and muscle damage markers in athletes under cumulative fatigue. Front Physiol 9:1137
Machin DR, Christmas KM, Chou TH, Hill, SC, Van Pelt DW, Trombold JR, Coyle EF (2014) Effects of differing dosages of pomegranate juice supplementation after eccentric exercise. J Physiol 271959
Mastaloudis A, Leonard SW, Traber MG (2001) Oxidative stress in athletes during extreme endurance exercise. Free Radic Biol Med 31:911–922
Matthaiou CM, Goutzourelas N, Stagos D, Sarafoglou E, Jamurtas A, Koulocheri SD, Haroutounian SA, Tsatsakis AM, Kouretas D (2014) Pomegranate juice consumption increases GSH levels and reduces lipid and protein oxidation in human blood. Food Chem Toxicol 73:1–6
Mazani M, Fard AS, Baghi AN (2014) Effect of pomegranate juice supplementation on matrix metalloproteinases 2 and 9 following exhaustive exercise in young healthy males. J Pak Med Assoc 64:785–790
Medved I, Brown MJ, Bjorksten AR, Murphy KT, Petersen AC, Sostaric S, Gong X, McKenna MJ (2004) N-acetylcysteine enhances muscle cysteine and glutathione availability and attenuates fatigue during prolonged exercise in endurance-trained individuals. J Appl Physiol 97:1477–1485
Mirończuk-Chodakowska I, Witkowska AM, Zujko ME (2018) Endogenous non- enzymatic antioxidants in the human body. Adv Med Sci 63(1):68–78
Miyazaki H, Oh-Ishi S, Ookawara T, Kizaki T, Toshinai K, Ha S, Haga S, Ji LL, Ohno H (2001) Strenuous endurance training in humans reduces oxidative stress following exhausting exercise. Eur J Appl Physiol 84(1–2):1–6
Morillas-Ruiz JM, Villegas García JA, López FJ, Vidal-Guevara ML, Zafrilla P (2006): effects of polyphenolic antioxidants on exercise-induced oxidative stress. Clin Nutr 25(3):444–453
Mrakic-Sposta S, Gussoni M, Moretti S, Pratali L, Giardini G, Tacchini P, Mastorci F, Borghini A, Montorsi M, Vezzoli A (2015) Effects of mountain ultra marathon running on ROS production and oxidative damage by micro-invasive analytic techniques. PLoS ONE 10:e0141780
Naghizadeh-Baghi A, Mazani M, Shadman-Fard A (2015) *Punica granatum* juice effects on oxidative stress in severe physical activity. Mater Socio Med 27:48–51
Nimse SB, Palb D (2015) Free radicals, natural antioxidants, and their reaction mechanism. RSC Adv 5:27986–280062
Otani H (2011) Oxidative stress as pathogenesis of cardiovascular risk associated with metabolic syndrome. Antioxid Redox Signal 15(7):1911–1926
Parker L, McGuckin TA, Leicht AS (2014) Influence of exercise intensity on systemic oxidative stress and antioxidant capacity. Clin Physiol Funct Imaging 34:377–383
Paschalis V, Theodorou AA, Margaritelis NV, Kyparos A, Nikolaidis MG (2018) N-acetylcysteine supplementation increases exercise performance and reduces oxidative stress only in individuals with low levels of glutathione. Free Radic Biol Med 115:288–297
Pinho RA, Silva LA, Pinho CA, Scheffer DL, Souza CT, Benetti M, Carvalho T, Dal-Pizzol F (2010) Oxidative stress and inflammatory parameters after an Ironman race. Clin J Sport Med 20:306–311

Ponce-Gonzalez JG, Corral-Pérez J, Sáez de Villarreal E, Vicente Gutierrez-Manzanedo J, De Castro-Maqueda G, Casals C (2021) Antioxidants markers of professional soccer players during the season and their relationship with competitive performance. J Hum Kinet 80:113–123

Powers SK, Deminice R, Ozdemir M, Yoshihara T, Bomkamp MP, Hyatt H (2020): Exercise-induced oxidative stress: friend or foe? J Sport Health Sci 9(5):415–425

Powers SK, Jackson MJ (2008) Exercise-induced oxidative stress: cellular mechanisms and impact on muscle force production. Physiol Rev 88:1243–1276

Powers SK, Nelson WB, Hudson MB (2011) Exercise-induced oxidative stress in humans: cause and consequences. Free Radic Biol Med 51(5):942–950

Reid MB (2016) Reactive oxygen species as agents of fatigue. Med Sci Sports Exerc 48:2239–2246

Roberts LJ II, Oates JA, Linton MF, Fazio S, Meador BP, Gross MD, Shyr Y, Morrow JD (2007) The relationship between dose of vitamin E and suppression of oxidative stress in humans. Free Radic Biol Med 43:1388–1393

Ryan MJ, Dudash HJ, Docherty M, Geronilla KB, Baker BA, Haff GG, Cutlip RG, Alway SE (2010) Vitamin E and C supplementation reduces oxidative stress, improves antioxidant enzymes and positive muscle work in chronically loaded muscles of aged rats. Exp Gerontol 45:882–895

Sadowska-Krępa E, Kłapcińska B, Kimsa E, Karpińsk R (2008) Effects of supplementation with red grape skin polyphenolic extract and interval swimming test on the blood antioxidant status in healthy men. Med Sport 12(1):1–7

Sanderson TH, Reynolds CA, Kumar R, Przyklenk K, Hüttemann M (2013) Molecular mechanisms of ischemia-reperfusion injury in brain: pivotal role of the mitochondrial membrane potential in reactive oxygen species generation. Mol Neurobiol 47:9–23

Santos S, Silva E, Caris A, Lira F, Tufik S, Dos Santos R (2016) Vitamin E supplementation inhibits muscle damage and inflammation after moderate exercise in hypoxia. J Hum Nutr Diet 29:516–522

Shadab M, Islam N, Khan Z, Khan F, Hossain MM (2014) Oxidative stress in sports persons after a bout of intense exercise: a cross sectional study. Biomed Res 25(3):387–390

Silva LA, Pinho CA, Silveira PC, Tuon T, De Souza CT, Dal-Pizzol F, Pinho RA (2010) Vitamin E supplementation decreases muscular and oxidative damage but not inflammatory response induced by eccentric contraction. J Physiol Sci 60:51

Skarpańska-Stejnborn A, Basta P, Pilaczyńska-Szcześniak Ł, Horoszkiewicz- Hassan M (2010) Black grape extract supplementation attenuates blood oxidative stress in response to acute exercise. Biol Sport 27(1):41–46

Spanidis Y, Stagos D, Orfanou M, Goutzourelas N, Bar-Or D, Spandidos D, Kouretas D (2017) Variations in oxidative stress levels in 3 days follow-up in ultramarathon mountain race athletes. J Strength Cond Res 31:582–594

Taghizadeh M, Malekian E, Memarzadeh Mr Mohammadi Aa Asemi Z (2016) Grape seed extract supplementation and the effects on the biomarkers of oxidative stress and metabolic profiles in female volleyball players: a randomized, double-blind, placebo- controlled clinical trial. Iran Red Crescent Med J 18(9):e31314

Toscano LT, Tavares RL, Toscano LT, Silva CS, Almeida AE, Biasoto AC, Gonçalves Mda C, Silva AS (2015) Potential ergogenic activity of grape juice in runners. Appl Physiol Nutr Metab 40(9):899–906

Tsang C, Wood G, Al-Dujaili E (2011) Pomegranate juice consumption influences urinary glucocorticoids, attenuates blood pressure and exercise-induced oxidative stress in healthy volunteers. Endocr Abs 25:139

Urquiaga I, Strobel P, Perez D, Martinez C, Cuevas A, Castillo O, Marshall G, Rozowski J, Leighton F (2010) Mediterranean diet and red wine protect against oxidative damage in young volunteers. Atherosclerosis 211:694–699

Valko M, Leibfritz D, Moncol J, Cronin MT, Mazur M, Telser J (2007) Free radicals and antioxidants in normal physiological functions and human disease. Int J Biochem Cell Biol 39:44–84

Vergeade A, Mulder P, Vendeville C, Ventura-Clapier R, Thuillez C, Monteil C (2012) Xanthine oxidase contributes to mitochondrial ros generation in an experimental model of cocaine-induced diastolic dysfunction. J Cardiovasc Pharmacol 60:538–543

Vezzoli A, Dellanoce C, Mrakic-Sposta S, Montorsi M, Moretti S Tonini A, Pratali L, Accinni R (2016) Oxidative stress assessment in response to ultraendurance exercise: thiols redox status and ROS production according to duration of a competitive race. Oxid Med Cell Longev 6439037

Wagner KH, Reichhold S, Hölzl C, Knasmüller S, Nics L, Meisel M, Neubauer O (2010) Well-trained, healthy triathletes experience no adverse health risks regarding oxidative stress and DNA damage by participating in an ultra-endurance event. Toxicology 278:211–216

Wiecek M, Szymura J, Maciejczyk M, Kantorowicz M, Szygula Z (2018) Anaerobic exercise-induced activation of antioxidant enzymes in the blood of women and men. Front Physiol 9:9

Zoppi CC, Hohl R, Silva FC, Lazarim FL, Neto JA, Stancanneli M, Macedo DV (2006) Vitamin C and e supplementation effects in professional soccer players under regular training. J Int Soc Sports Nutr 3:37

Chapter 12
Comparison of Selected Biomechanical and Physiological Variables Among Novice and Experienced University Basketball Athletes

Ridham Thakral, Anilendu Pramanik, and Harish Kumar

1 Introduction

Basketball game involves ball handling, dribbling, passing, shooting and defense; each team has ultimate aim to shoot the ball through the opponents' basket. It is played by two opposing teams consist of 5 playing plays each on a court measuring 29 × 15 m. Shooting may be considered the most important technique in the game. To increase the score player as to convert many shots in opposite court, therefore a variety of shots are developed which included the hook shot, the jump shot and the layup shot.

Basketball is an aerobic-based anaerobic sport which requires high-intensity activities such as jumping (for rebounds, blocks and shots), turns, dribbles, sprints, screens and low-intensity activities such as walking, stopping and jogging. Frequent stoppages in games allow players to recover between bouts of activity, thus allowing repeated high-intensity game play (Drinkwater, 2008). Aerobic capacity is positively associated with recovery during repeated high-intensity bouts. Moreover, the high-intensity movements of basketball players are closely related to the development of strength, speed and agility (Castagna et al. 2007; Hedrick 1993; Meckell et al. 2009). During a basketball game, professional players cover about 3500–5000 m (Janeira and Maia 1998). Each player performs about 1000, mainly short, activities lasting around 2 s; time motion analysis has shown that these short activities are

R. Thakral (✉) · A. Pramanik · H. Kumar
Department of Sports Science, Punjabi University, Patiala, Punjab 147002, India
e-mail: ridhamthakral1998@gmail.com

A. Pramanik
e-mail: anilendu.myas@gndu.ac.in

H. Kumar
e-mail: harish_ss@pbi.ac.in

A. D. Singh et al. (eds.), *Sports Ergonomics*, Design Science and Innovation,
https://doi.org/10.1007/978-981-97-7804-1_13

performed with a different frequency according to the player's position (Abdelkrim et al. 2007). Explosive strength, take-off power, speed and agility are abilities that make an important contribution to efficient movement with and without the ball, thus play an important role in basketball technique and tactics (Erculj et al. 2010). The level of these abilities, that is, the motor potential, is most often measured using various motor tests with and without the ball (Colli et al. 1987). In basketball practice, motor tests are the most suitable and applicable because they are implemented in conditions similar to those of training or competition (Erculj et al. 2010).

A performance level can only be increased by working and training on all major components, i.e. physical fitness, technique, tactics, physiological qualities, coordination and psychological qualities. These include the high skill level, flexibility, muscular strength, speed, agility and importantly the specific use of different physiological, body composition and biomechanical variables (Alemdaroğlu, 2012). Both variables testing is important in the game to identify potential talent in the player and for his or her selection. Testing also helps to decide which training programs player should follow. Therefore, together these variables will provide a perfect combination to ensure analysis of every physical component of the game.

I have examined shoulder strength and power are frequently using the isokinetic shoulder joint test and the vertical jump test; no studies have focused on the correlation between performance in these two motor skills. These scattered findings may be due to a number of differences, such as joint angular velocities and the positioning of the participants, affecting muscle length and velocity of contraction, participant characteristics and methods used for calculation strength of joint in isokinetic dynamometry.

Some studies have attempted to correlate the results of the WAnT and isokinetic tests. Arslan (2005) found that there were significant correlations between explosive leg strength and anaerobic performance. It was reported earlier that isokinetic concentric shoulder extension strength was significantly correlated with peak and mean power at all contraction velocities (60°, 90°, 120°). However, there was significant correlation only between 90∘ shoulder flexion strength and peak power for isokinetic concentric shoulder flexion.

While some studies have investigated the relationships between isokinetic knee strength, anaerobic performance, sprinting ability, agility and vertical jump performance in other athletes, an insufficient number of studies have been conducted on basketball players (Tavino et al. 1995). Therefore, the purpose of this study was to investigate the relationship between isokinetic shoulder strength, anaerobic performance, sprinting ability, agility and vertical jump performance in both trained and untrained basketball players.

As basketball sport is the most trending sport in India as well as around the world but still there are many limited researches on this game whereas if we talk about the research on combination of both biomechanical and physiological variables of basketball players then that is also limited. That is why I have decided to research on the combination of these variables. It is hypothesis that the combined effect of both the variables (physiological as well biomechanical variables) will enhance the performance of basketball players.

2 Design and Methodology

Ethical clearance was taken from university ethical committee. Informed consent was taken from subjects.

It is an experimental study consisting of 11 basketball players which were divided into 2 groups (experimental group, $n = 6$ and control group, $n = 5$), experimental group consists of at least 1 year of experience, while control group consists of minimum 5 years of experience.

3 Data Collection

The data was collected during the period of February 2021 to April 2021. During the commencement of the data collection, all of them were participated for basal data collection in following sequence: -

- Anaerobic Data: Device [Monarch Bicycle Ergometer] with 30 s Wingate 30 s.
- Aerobic Data: Device [Metamax 3B] with Bruce protocol.
- Forced platform: Device [Kistler's Forced Platform (QUARTTRO)] by checking maximum effort jump with both legs (freehand).
- Strength: Device [Strength Analysis System (BIODEX)] by collecting only shoulders strength.

Before data acquisition, all participants had informed the benefits and risk of the instruments and a brief description of the study purpose were provided to them. Thereafter, a signed consent was obtained from each of them. The subjects were also given full awareness about the significance of the study.

4 Result

We have evaluated a selected parameter of trained and untrained basketball athletes. We also measured their selected anaerobic parameter like peak power, average power, minimum power, power drop and total energy produced during 30 s Wingate anaerobic test. We also evaluated the counter movement jump force parameter by using Kistler's forced platform (QUARTTRO) and studied the following selected variables, the variables are relative maximal force, relative maximal force during counter movement, average, vertical take-off velocity, average power, average force, average velocity, jump time, counter movement time, flight time, push off flight, eccentric deceleration phase flight, fat peak concentric and active movement time. In addition, we also measured the aerobic parameter by Bruce protocol on an electronical bark treadmill. We observed heart rate, $V'o_2$/kg, V'E, METS, $V'Co_2$ and VD/VT. All these scaled variables satisfied the normal distribution and therefore further analysis took place (Tables 1, 2, and 3, 4 and Fig. 1).

Firstly, we performed a correlation analysis among those variables, and we prepared a correlation matrix, which is given in below table. Then we were interested to find the mean differences of all these variables among the untrained and trained participants. Surprisingly curtain variables, e.g., total energy of anaerobic variables

Table 1 Mean and standard deviation of force platform data

Variables	Control group		Experimental group	
	Mean	Standard deviation	Mean	Standard deviation
General parameters				
Relative maximal *F* [%BW]	226.25	17.55	203.77	27.68
Relative maximal *F* during counter movement [%BW]	191.28	41.05	167.88	31.45
A [m/s^2]	2.07	0.55	2.24	0.87
Vertical take-off *V* [m/s]	2.27	0.31	1.82	0.89
Average *P* [W]	1361.69	230.98	1161.78	145.44
Average *F* [N]	1060.86	140.77	1148.68	137.11
Average *V* [m/s]	1.17	0.24	1.26	0.22
Time parameters				
Jump *T* [s]	1.04	0.31	1.04	0.29
Counter movement *T* [s]	0.67	0.25	0.62	0.28
Flight *T* [s]	0.47	0.06	0.47	0.08
Force impulse parameters				
Push off *FI* [Ns]	121.91	69.88	152.27	29.17
Eccentric deceleration phase *FI* [Ns]	55.98	28.46	48.84	32.58
Fat peak concentric [N/kg]	20.78	0.52	19.99	2.71
Active movement *T* [s]	0.74	0.42	0.83	0.26

Table 2 Mean and standard deviation of anaerobic data

Variables	Control group		Experimental group	
	Mean	Standard deviation	Mean	Standard deviation
Peak power [W/kg]	8.00	1.07	7.71	1.31
Average power [W/kg]	6.01	0.49	5.57	0.76
Minimum power [W/kg]	2.75	0.94	2.66	1.41
Power drop [%]	66.20	9.20	66.72	16.67
Total energy produced [J]	13,002.40	2274.93	11,190.67	1694.16

Table 3 Mean and standard deviation of shoulder strength data

	Right				Left			
Variables	Control group		Experimental group		Control group		Experimental group	
	Mean	Standard deviation	Mean	Standard deviation	Mean	Standard deviation	Mean	Standard deviation
EXT rotation 60 deg/sec								
Peak torque	24.46	5.51	21.78	6.90	21.74	5.17	26.22	9.08
Peak torque/ body weight	32.60	6.34	31.00	9.09	29.42	5.42	33.63	10.27
Average peak torque	18.72	3.10	18.87	7.06	17.92	6.04	21.75	4.69
EXT rotation 90 deg/sec								
Peak torque	16.86	6.68	22.60	4.02	27.22	9.28	24.40	8.60
Peak torque/ body weight	22.10	7.91	28.98	8.11	28.68	9.83	25.42	5.31
Average peak torque	17.18	6.32	18.93	3.95	21.38	10.28	20.98	8.55
EXT rotation 120 deg/sec								
Peak torque	19.04	7.32	20.15	5.28	21.84	4.67	21.63	8.29
Peak torque/ Body weight	25.66	7.17	19.38	5.83	28.84	6.78	27.2	6.88
Average peak torque	19.22	6.92	15.12	6.26	20.18	4.13	17.88	6.27
INT rotation 60 deg/sec								
Peak torque	42.70	8.46	48.07	6.80	38.34	4.83	40.43	7.22
Peak torque/ body weight	48.26	8.12	45.78	9.39	45.48	10.67	39.53	4.87
Averagepeak torque	38.34	5.50	42.65	4.33	29.14	4.09	34.05	6.06
INT rotation 90 deg/sec								
Peak torque	47.40	9.94	45.25	6.64	43.80	11.36	43.57	7.57
Peak torque/ body weight	51.68	5.07	49.43	8.33	40.32	11.50	39.30	5.92
Average peak torque	45.08	5.72	40.95	9.49	41.42	11.12	38.75	6.51
INT rotation 120 deg/sec								
Peak torque	42.76	9.97	41.10	5.13	46.66	6.70	49.55	8.07
Peak torque/ body weight	39.64	11.15	39.07	7.71	41.44	9.02	45.02	9.02
Average peak torque	40.28	9.65	38.78	5.47	39.58	11.74	43.33	11.74

Table 4 Mean and standard deviation of vo2 max data

Variables	Control group		Experimental group	
	Mean	Standard deviation	Mean	Standard deviation
V'o2/Kg	44.80	2.39	40.17	4.54
HR	184.20	6.94	194.33	4.59
V'E	115.28	7.03	120.60	5.50
METS	12.76	0.75	11.47	1.27
V'Co2	3.55	0.72	3.01	0.50
VD/VT(est)	0.10	0.04	0.11	0.03

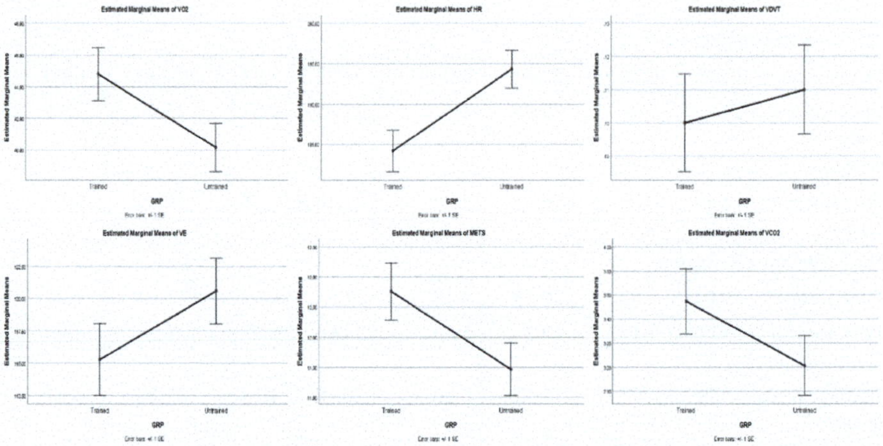

Fig. 1 Metabolic parameters changes are shown in two different group of studied population

and heart rate of aerobic variables showed significant differences ($T = 2.3$, $DF = 9$, $P = 0.046$); ($T = -2.91$, $DF = 9$, $P = 0.017$), respectively. Further, it is to be noted that the remaining variables also expressed some change, some difference within trained and untrained groups but significance was not evident.

As we have evaluated anaerobic, force and aerobic data within short time courses of data collection. We were very much interested to find out the factor analysis of our scaled variable. We prepared a scree plot and can predict which variable are important for determining performance of basketball athletes. Finally, we analyzed a liner multiple regression to report the aerobic variable along with other variables. From the data, we have observed that energy expenditure and minimum power are important predictors to report the dependent variable. All the variables' differences among trained and untrained group are shown below picture (Fig. 2).

All participants in our study subjective to strength evaluation to Biodex system. As they were basketball players, we evaluated their shoulder external and internal rotation strength for 3 angles (60°, 90° and 120°) on left and right side of their body. So, we performed a four-way ANOVO of all the strength related variables.

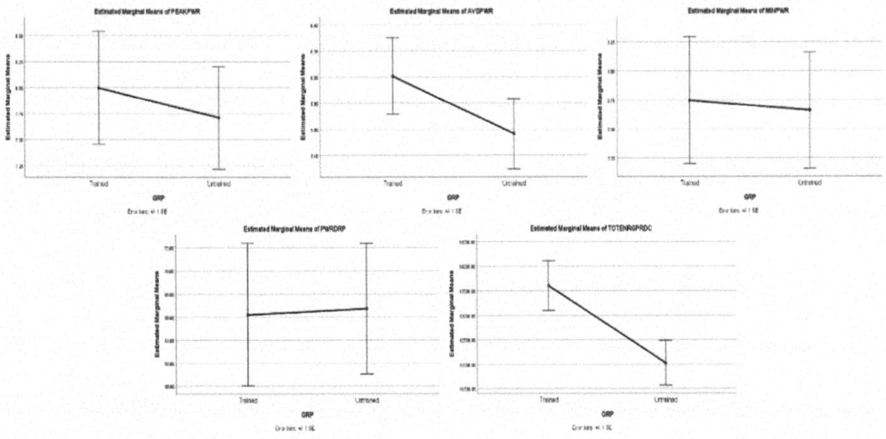

Fig. 2 Anaerobic variables change in two different groups of participants in studied population

The descripted statistics of their variables are shown in following table from the result, we have observed that only rotation has significant effect [$F = 281.97$, DF = (1108) and $P < 0.001$]. All most seventy-two (72°) percent of studied population had affected by these rotational variables other independent factor, e.g., side, angle, group did not show any significant effect on the dependent variables. Pair-wise comparison by Bonferroni post-hoc test revealed that internal rotation significantly shown higher value then their opposite counterpart. The graphical representation of all these dependent variables changes within the experimental conditions is being shown here (Figs. 3, 4).

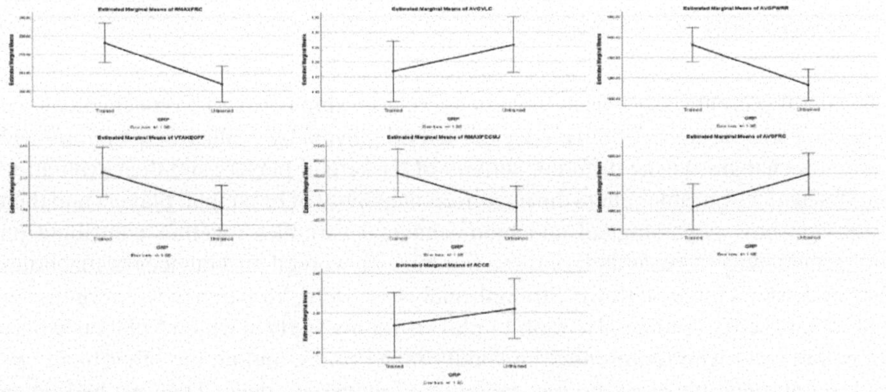

Fig. 3 Kistler's force platform-based data changes in different group of studied populations

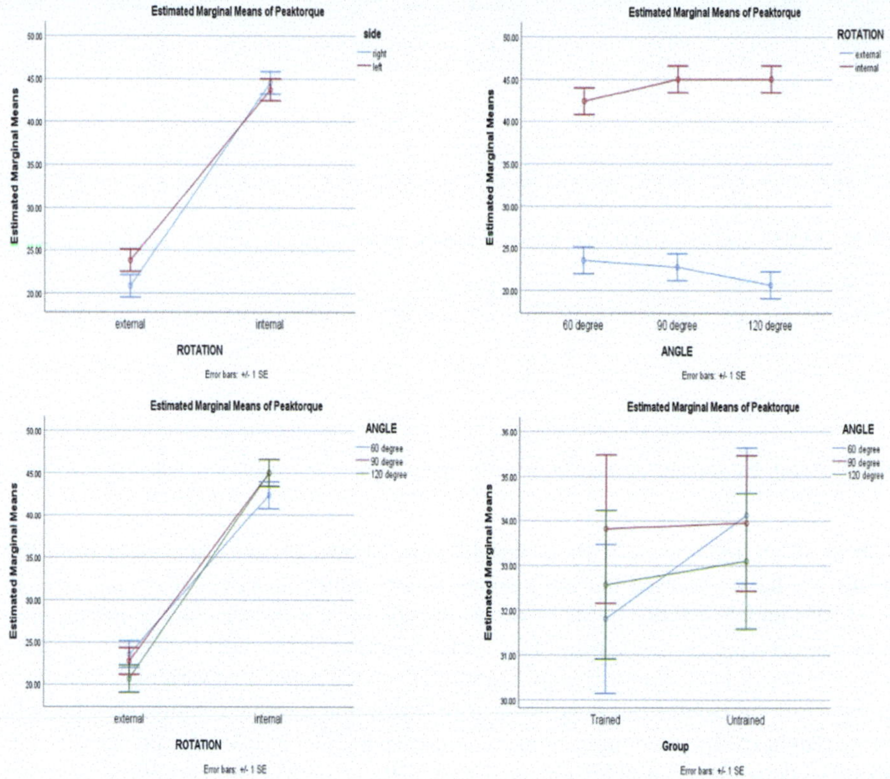

Fig. 4 Upper shoulder strength variables among different groups of studied population

5 Discussion

The aim of present study was to find out the relationship between biomechanical and physiological variables of university basketball players by evaluating physiological variables changes during different activity of basketball players and then evaluating biomechanical variables alters during different activity of basketball players and then observing how physiological and biomechanical variables together contribute on different activity of basketball players. For this, we worked on 4 devices or machines that are Kisler's force platform, Strength analysis system, Monarch bicycle ergometer and Metamax 3B. And we also took 11 players for my study in which 5 players where trained players who represented university many times, and on the other hand, we took 6 untrained players who just represented university once. Then we had taken every player's one-time data on each device. I used strength analysis system to find out the shoulder strength of both hands of the player as shoulder strength is used to shoot the basket as well as used to do dribbling or while passing. To calculate the anaerobic capacity of the player, we used monarch bicycle ergometer as anaerobic capacity is the capacity in which the person run so fast that he or she doesn't even get the time to take oxygen. And anaerobic capacity is required by the player while doing fast break. Then the Kisler's force platform was used to calculate the force or power exerted by the body while landing on the ground from a free hand jump. Then the last device which I used was Metamax 3B, this device was used to calculate the lung capacity of the player or we can say this device was used to calculate how much the player's body release carbon dioxide and take oxygen while doing workout. By all these data, we found out that the trained players are more fit and active as compared to untrained players. As trained players do training or do workout daily that is why all trained players data is good as compared to untrained once, whereas untrained players are irregular toward their practice, so their body data lacks. If untrained players start doing workout daily or start doing training daily, then they will also get fit and body data will also improve and their game skill will also get improved.

6 Conclusion

We have observed athletic and non-athletic group comparison on selected parameters like Kistler's force platform, anaerobic, aerobic and strength variables. Conventionally, athletic group showed improved response. Thus, above knowledge could be beneficial for coaches and management staff of basketball game and a scientific database could be generated from this study.

References

Alemdaroğlu U (2012) The relationship between muscle strength, anaerobic performance, agility, sprint ability and vertical jump performance in professional basketball players. J Hum Kinet 31(1):149–158

Bale P (1986) A review of the physique and performance qualities characteristic of games players in specific positions on the field of play. J Sports Med Phys Fitness 26:109–122

Ben Abdelkrim N, El Fazaa S, El Ati J (2006) Time-motion analysis and physiological data of elite under- 19 Basketball players during competition. Br J Sports Med 41:69–75

Bergh U, Sjodin B, Forsberg A, Svedenhag J (1991) The relationship between body mass and oxygen uptake during running in humans. Med Sci Sports Exerc 23:205–211

Bosco C, Luhtanen P, Komi PV (1983) A simple method for measurement of mechanical power in jumping. Eur J Appl Physiol 50:273–282

Castagna C, Chaouachi A, Rampinini E, Chamari K, Impellizzeri F (2009) Aerobic and explosive power performance of elite Italian regional- level basketball players. The J Strength Conditioning Research 23(7):1982–1987

Castagna C Vincenzo MM D'Ottavio S Annino G Padua E and Bishop D (2007). Relation between maximal aerobic power and the ability to repeat sprints in young basketball players. J Strength Cond Res 21:1172–1176

Chaouachi A, Brughelli M, Chamari K, Levin GT, Abdelkrim NB, Laurencelle L, Castagna C (2009) Lower limb maximal dynamic strength and agility determinants in elite basketball players. The J Strength Cond Res 23(5):1570–1577

Church JB, Wiggins MS, Moode FM, Crist R (2001) Effect of warmup and flexibility treatments on vertical jump performance. J Strength Cond Res 15:332–336

Delextrat A, Cohen D (2008) Physiological testing of basketball players: toward a standard evaluation of anaerobic fitness. The J Strength Cond Res 22(4):1066–1072

Eston RG, Reilly T (2001) Kinanthropometry and Exercise Physiology Laboratory Manual: Tests, Procedures and Data: Anthropometry, vol 1. Routledge, London

Hakkinen K (1988) Effects of the competitive season on physical fitness profile in elite basketball players. J Human Mov Stud 15: 119–128.

Hoare DG (2000) Predicting success in junior elite basketball players—the contribution of anthropometic and physiological attributes. J Sci Med Sport 3(4):391–405

Hoffman JR, Maresh CM (2000) Physiology of basketball. In: Garrett WE, Kirkendall DT (eds) Exercise: Basic and Applied Science. Lippincott Williams & Wilkins, Baltimore, MD, pp 733–744

Hoffman JR, Stavsky H, Falk B (1995) The effect of water restriction on anaerobic power and vertical jumping height in basketball players. Int J Sports Med 16:214–218

Hoffman JR, Tenenbaum G, Maresh CM, Kreamer WJ (1996) Relationship between athletic performance tests and playing time in elite college basketball players. J Strength Cond Res 10:67–71

Hoffman JR, Epstein S, Einbinder M, Weinstein Y (1999) The influence of aerobic capacity on anaerobic performance and recovery indices in basketball players. J Strength Cond Res 13:407–411

Jakovljević S, Karalejić M, Pajić Z, Gardašević B, Mandić R (2011) The influence of anthropometric characteristics on the agility abilities of 14 year-old elite male basketball players. Facta Universitatis-Series: Physical Education and Sport 9(2):141–149

McInnes SE, Carlson JS, Jones CJ, McKenna MJ (1995) The physiological load imposed on basketball players during competition. J Sports Sci 13:387–397

Tavino LP, Bowers CJ, Archer CB (1995) Effects of basketball on aerobic capacity, anaerobic capacity, and body composition of male college players. The J Strength and Cond Res 9(2):75–77

SPRINGER NATURE

GPSR Compliance

The European Union's (EU) General Product Safety Regulation (GPSR) is a set of rules that requires consumer products to be safe and our obligations to ensure this.

If you have any concerns about our products, you can contact us on ProductSafety@springernature.com

In case Publisher is established outside the EU, the EU authorized representative is:

Springer Nature Customer Service Center GmbH
Europaplatz 3
69115 Heidelberg, Germany

The manufacturer's authorised representative in the EU is Springer Nature Customer Service Centre GmbH, Europaplatz 3, 69115 Heidelberg, Germany. If you have any concerns regarding our products, please contact ProductSafety@springernature.com

Printed and bound by CPI Group (UK) Ltd, Croydon, CR0 4YY
01/10/2026
02225980-0005